Nature, Change
and the Human Endeavor

Seeking Answers

Richard Revel

DETSELIG
ENTERPRISES LTD

Nature, Change and the Human Endeavor: Seeking Answers

Library and Archives Canada Cataloguing in Publication

Revel, Richard David, 1946-
     Nature, change and the human endeavor: seeking answers/Richard Revel
ISBN 1-55059-263-7
     1. Global environmental change. 2. Ecosystem management. 3. Human ecology. 4. Nature–Effect of human beings on. I. Title

GF41.R49 2004   304.2'8   C2004-904253-X

Detselig Enterprises Ltd.
210, 1220 Kensington Road NW
Calgary, Alberta T2N 3P5

Phone: (403) 283-0900
Fax: (403) 283-6947
Email: temeron@telusplanet.net
www.temerondetselig.com

We acknowledge the support of the Government of Canada through the Book Publishing Industry Development Program (BPIDP) for our publishing program.

We also acknowledge the support of the Alberta Foundation for the Arts for our publishing program.

Alberta Foundation for the Arts

Alberta COMMUNITY DEVELOPMENT

COMMITTED TO THE DEVELOPMENT OF CULTURE AND THE ARTS

ISBN 1-55059-263-7
SAN 113-0234
Printed in Canada

Cover photograph by Avril Revel
Cover design by Alvin Choong

Dedication

To the 'Via Media' and all those who, with humility, and in the face of uncertainty and change, struggle to seek a prudent balance between conservation and development in accord with the common weal.

Table of Contents

Acknowledgements

Acknowledgements are often a difficult and tricky thing to write, particularly so in this case as the book covers much more than my objective observations on nature. It ventures into the arena of values and ethics which of course reflect my life experience. I acknowledge, with much appreciation, all those who have shaped and molded my journey thus far – it has been a very rich and wonderful one.

I specifically want to acknowledge two professors who played significant roles in opening my eyes to the wonders of nature; Dr. Lochan Bakshi who remains a good friend and Dr. Vladamir J. Krajina, now deceased. These men gave of themselves unstintingly and I owe them an enormous debt of gratitude.

The University of Calgary and the Faculty of Environmental Design have offered me a rich hospice for nearly thirty years where my colleagues and graduate students have debated and challenged me. Universities are indeed intellectually fertile places – Environmental Design has been particularly rich as its denizens explore at the integrative margins of science, design, art and values in order to seek appropriate human interventions in the natural and built environments. I have been privileged to have fine colleagues and students.

Dr. Roland Priddle, retired Chairman of the National Energy Board of Canada, initiated me into the world of regulatory tribunals and provided a wonderful example of fairness and balance in decision taking in the public interest. We are indeed blessed as Canadians to have people such as him at the helm of our regulatory institutions. Roland also very kindly read and commented on the draft manuscript. My time with the NEB fundamentally changed my perspective on environmental management and decision making, and made me a skeptic of much of what I hear and read. It has led me to begin each of my courses with the following admonition to students:

> Never believe what I, or anyone, tells you unless it makes sense, is well supported by evidence, is reasonable or is consistent with your experience – I could be wrong, deceiving you or simply ignorant. Challenge everything and always look for the contrary evidence, as well as the motives, to claims people and institutions make. Evidence and argument are not the same, and bias and self-interest are indeed part of the human condition.

I must acknowledge Graham McDonald who reviewed and provided much editorial advice and encouragement. Graham's generosity and encouragement were instrumental in my endeavor.

My appreciation also goes out to Detselig Enterprises and Kim Robertson, my editor. Kim, thanks to her well-honed skills, has not bludgeoned me too badly, has been a joy to work with and has offered much to you, the reader, by way of excising the rubbish while still leaving the story.

I owe much to my parents Ethna and David, and to my siblings, Rhoda, John, and Bert, and their spouses, George, Shae, and Elizabeth.

Finally I have been richly blessed and also owe much to my own little family. Avril, my wife, and my daughters Andrea and Fiona. These three have done much to shape my journey – they have been my most ferocious critics, best friends and generous companions.

Avril has travelled at my side through most of the journey and provided another perspective on the world based on her gender and her background as a medic. We have shared marvelous discussions and a rich life. She also willingly listened and critiqued my written offerings – she is generous. Thank you, Avril.

Any errors or omissions fall squarely on my shoulders.

Introduction

Nature! Humans! Who is doing what to whom in this dance between humans and the rest of nature? When? Where? How? Why? And for that matter – Why Not? We humans have an interesting relationship with our environment.

Somewhere along the continuum of life on Earth, amidst the ongoing changing of species as the grand experiment of life continues and species come and go like actors on a stage, a rather interesting species began to develop. This species noticed there was a connection between its actions and its well-being. It was self-conscious. It noticed. It enquired. It pondered relationships among things. It asked questions, reflected, imagined and answered its own queries. It, this interesting species, still asks questions and tries to provide answers.

When one writes on such broad issues as the title of this book implies, they expose themselves to assertions of bias and criticism. These cries are largely valid because each writer and each critic is the sum total of their experiences and individual experiences are always different unless the writer and the critic are one and the same.

Such challenges and criticisms are entirely appropriate. I welcome them and make no apologies for my analyses. After all this is my story. My analyses of humans and nature will, I hope, be judged on their own merit; preferably in the arena that combines reason, evidence, experience, probability and curiosity and not in the arena of theology. When one holds a belief, regardless whether it concerns God or the effects of humans on the environment there is unequivocally no room for discussion unless the holder is prepared to set the belief aside, at least temporarily, and examine that belief objectively. There must at least be some level of methical doubt or we cannot communicate.

I acknowledge that in areas where human values and the human condition come into play there is a huge body of carefully crafted research and literature dealing with values, attitudes, beliefs, art, beauty and theology. These shared values are all important as they get translated into public policy and laws related to conservation and development, management practices and codes of acceptability of human conduct and interchange. Hopefully they do, for the human endeavor requires some measure of consensus if a reasonably civil and caring society is to result. Blatant self-interest doesn't work well among social animals.

I also acknowledge there is no human endeavor that is value free: science, philosophy, city planning, engineering, medicine or mathematics all have their foundation in human values and to claim otherwise would be to deny the basic humanity of our species. We all bring bias to our views and those biases are important in understanding thoughts and arguments.

As to my own biases, I offer the following summary, believing it to be relevant to any judgments that you, the reader, are going to make on the merits of my various positions. I am a plant ecologist by formal education, a professional biologist by registration and a professor of environmental science. I am of an interdisciplinary persuasion and have a proclivity for taking the fruits of deductive research and building big pictures. In short, I am a synecologist. My interests and research have focused on various aspects of environmental planning, management, development and conservation. I admit a bias toward concrete solutions and techniques that demonstrably improve current environmental practices and which are testable in the field. I consider myself an applied field ecologist rather than a theoretical ecologist.

I am a skeptic of purported grand solutions. I hold that natural systems have, over a long history, demonstrated their ability to endure and adjust to major environmental changes and assaults of grand proportions. Indeed, it seems to me that organisms have met each major environmental change with brilliant new solutions. Furthermore my observations suggest that when humans cease, alter or reduce their interventions on the environment, that environment, through its self-regulating ability, will establish ecological systems that are entirely appropriate to the prevailing conditions. I acknowledge that both the conditions and the systems will change and am not terribly troubled by this fact. I am also quite happy to decide that I value a particular system and I am willing to actively intervene to prevent natural changes from taking place in order to maintain what I value. I might be classified as reserved technological optimist as many technological innovations have improved our ability to meet our ends through either increased efficiencies or reduced environmental disturbances – but technology has its problems. We humans have found resources where other species have not and the more we use these, the more we leave for the other species. I consider we humans as part of nature, albeit an adaptable, manipulative and powerful species, that is ultimately subject to the limitations of all natural laws. Our genetics make us this way.

Finally, in declaring my biases, I have been involved in regulatory decision making in both Canada and South America. I have had both the pleasure and the responsibility of sitting as a Member of the National Energy Board of Canada and have been charged to consider energy related projects under the Canadian Environmental Assessment Act and the National Energy Board Act. In these positions I have been, along with my colleagues, charged with making decisions based on the concepts of *likely environmental impacts that are not justifiable, public convenience and necessity and the public good.* While I no longer hold the above mentioned positions, I am a great believer in the concept of public service, the pubic good and I hold that civil society is essential to freedom and justice.

I have had the pleasure of many years studying and enjoying our marvelous world out of sheer curiosity and wonderment and I love it dearly. I marvel at the diversity of earth's millions of life forms, their adaptations to both stable and changing environments and nature's ability, in the face of massive environmental changes and assaults, to re-cloak the earth with an ever increasing and changing abundance of organisms. This diversity is primarily due to, what seems to me to be a fact, imperfection; in the genetic code and in the atoms of elements from which all matter is constructed. I have concluded (conveniently so my wife says) that imperfection is the only path to ecological and social richness and that perfection is the ultimate path to failure of life on earth.

I have also taught and debated the topics in this book with more than a few bright graduate students over the past 35 years who have molded, challenged and reshaped my thoughts. Life has been rich for me thanks in no small part to their challenges. I have immense regard for humans with their curiosity, their sense of the past and their concept of the future. I suspect that no other species operates on such a plane of consciousness although from a careful epistemological perspective this claim is unprovable because only we can define the test criteria and only we can judge the outcome.

Perhaps my thoughts will spur others to examine their beliefs and values about the environment and possibly even alter them in their pursuit of understanding relationships among humans, technology, the environment and our grand role in the scene of life and human endeavor. Perhaps others will examine my arguments and show me to be wrong. If that is the case, then I benefit for indeed humility is an important teacher in life and to be shown to be wrong a significant honor. Unquestioning acceptance of ideas is akin to either ignorance or indifference. It is my hope that I can build arguments to show that in the grand scale of life humans are relatively unimportant in the well being of nature; life will unfold without us quite nicely. The question of Cardinal Newman is interesting- "If a tree falls in the forest and nobody hears it, does it exist?" In true Darwinian form "At the banquet table of nature there are no invited guests. You get what you can take and you keep what you can hold"! I will add to this"... but friends and supporters help." When a species can no longer hold its spot on the face of the earth it is eliminated and replaced.

Finally, I sincerely hope that I can provide sound arguments in support of careful development and active conservation efforts. I am convinced that humans, in order to protect their future for the long term require better education, more technology so that we can be reduced in our want, meet our basic needs and cause less human induced environmental change that risks our species – for our own well being! Nature can take care of itself; we in contrast, will be the ultimate recipients of the laws of nature if we fail to attend to careful environmental management.

When the environment changes so that we can no longer adapt to the new conditions we will be replaced. I have never seen an ecosystem without integrity, an unhealthy ecosystem or an area of the world where fundamental ecological processes are not functioning. I have certainly seen lots of ecological systems that do change and in some cases I would have preferred they had not. The problem in all such cases was mine – not nature's. With this introduction over and my biases, basic premises and starting points stated let me get on with the story. I hope you enjoy the read.

On a Small Private Garden

I am an avid gardener, not a particularly meticulous one but an avid one nonetheless. I am not one of those careful gardeners who try to push the limits of climatic tolerance in order to get palm trees to grow in arctic climates. My hedges are not perfectly trimmed nor are the beds weed free. I do my best on a quarter of an acre to keep things modestly under control so as not to offend the sensibilities of my family and neighbors.

About fifteen years ago my family and I bought this lot and built a home on it. Northwest Calgary is not an area whose climate is noted for its kindness to either gardens or gardeners. The gardener's life is a struggle. The lot was flat and I proceeded to reconfigure it and create hills, a pond, forests and grasslands. The flat became undulating with slopes of different angles facing north, south, east and west and the degree of topsoil varying from a few centimeters to a few feet. I also built a sunken garden of about 500 square feet in area with two terraces for stability to a depth of about four feet so that we could enjoy protection from the elements and also gain access to the outside from our developed basement.

The elements of garden design were those so beautifully articulated by the noted English landscape gardener Sylvia Crowe: unity, scale, time, space division, light and shade, texture, tone, color and style. These can be achieved with various methods of execution involving a host of other principles related to shape, form, line, perspective, harmony and motion. Human notions all of them. These principles were altered and guided by some of the principles articulated by Chinese and Japanese masters who so carefully developed techniques to translate two-dimensional scroll paintings into three-dimensional gardens. My goal was to develop a contemplative garden where I could relax, think and observe the abundant changes that take place in nature on scales ranging from hours, to days, to seasons to years. In short I wanted a place to watch what nature did over time yet I did not want an eclectic tacking on of various elements of different cultures and styles; it had to reflect and be true to me and my curiosities, true to the local landscape in miniature, and suitable for the needs of our family. Private garden design is first and foremost a private and personal act, an expression of one's core; only secondarily is it a public act. If there is truth in the personal design then inevitably one's friends and intimates will value and enjoy it.

I planted shrubs, trees, and many perennial plants – not in neat rows but in areas that conformed to my design concept. I also seeded a lawn but not with great care. While some of my neighbors labored long and hard to get their land perfectly level before seeding, seeded it with care and tended it endlessly until the young grass grew, I simply rough-leveled it to my approximate specification, left medium sized clods of earth on the surface and seeded it abundantly before heading off for a month's holiday. The whole lawn process took little more than a morning. When I returned the grass had a good catch though not uniform; the clods had been broken-down by the rain and the lawn was quite level. The clods had afforded protection from drought for the young seeds as they attempted to get their roots into the wet soil after germination.

In case you are thinking the garden is a masterpiece of brilliant design to be numbered among the world's great gardens let me assure you that it is not. It is however diverse, has a lot of private spots to take refuge in, has hosted a lot of social gatherings and our cat liked it. She seldom strayed far from our yard and had 'cat nests' and resting spots throughout; in the native rough fescue grassland, under shrubs, on the hills and beside the pond. She was able to chase butterflies, bugs and in her younger days, birds. In her latter years, she was too old to chase birds but she lay in her cat nests, watched the birds and dreamed of her youth.

On a Garden Pond:
Lessons in Change and Stability

Rather than attempt to garner praise for my meagre gardening skills, I have used my own garden to provide a foundation from which to springboard into a story of ecological change and stability. You see, one of the secrets of gardens with life is water. Water in a garden adds motion and constant change as light conditions shift, clouds pass overhead or the sun shines and reflects off it. I built a pond; a small oval shaped lake about 18 feet across and 30 feet long. For safety reasons it is only about six or eight inches deep.

The earth was excavated and a used plastic swimming pool liner placed in the hole. This was then covered over with cobbles between one half and three inches in diameter and a few large boulders were placed strategically throughout. I had no grand aspirations of designing the ultimate water garden masterpiece. My goal was much less inspired. I wanted to see what nature would do with this wet spot that I had created. The pond is not elaborate, has no water re-circulation, filtering or elaborate maintenance. The only water it receives is from rain, snow or the garden hose when I remember. It has got dry at times over the years but it has prevailed.

There are many natural ponds in northwest Calgary left as a result of glacial actions and ablating ice; the type of depressions formed when blocks of ice from the melting glaciers are surrounded by till or outwash and the melting of the ice leaves a depressional void. These natural depressions collect runoff water and form ponds that are truly wonderful places with an abundance of plants and animals both large and small. Ducks and geese breed there, muskrat abound. They are surrounded by aspen copses and native grasslands, some of which are used as horse and cattle pasture. It has become an attractive and popular area for small acreage developments for those people with a proclivity to living the country life near the city. The owners are generally quite rich and they guard and protect their little pieces of nature with passion. The ponds and native grasslands are safe and well-protected thanks to wealth.

To kick-start my garden pond I visited one of these natural ponds with a bucket and shovel and dug up some of the bottom muck along with a few *Carex*, *Scirpus* and *Equisetum* plants. Nothing elaborate, just one bucket full. This muck I placed in five small locations around my garden pond. I have watched and enjoyed that pond for about fifteen years now and I will describe in general terms what changes have taken place. I won't offer a careful scientific description or give a long academic diatribe on the merits of ponds in the greater scheme of nature conservation and species protection. Suffice it to note that my observations have led me to supervise two wonderful graduate students who did their theses on matters related to wetlands and their role in management of urban storm water management. It has also played a role in my involvement in experimental work, along with some good colleagues, dealing with changes in practice of Calgary's storm-water management. I do not claim it to be the parent of current practice but it certainly sparked my enthusiasm for the incorporation of ecotechnologies in storm-water management. Calgary is currently very active in this area of management. Nature cleanses itself quite well and we can learn a lot, and benefit by, copying her processes.

During the early summer of 1989, the first year of my pond, the system was unstable and did not have a huge abundance of life. It was a bit like a child's wading pool covered with cobble in the bottom. I spent hundreds of hours sitting in a chair and watching it. Gradually the plants from the muck began to grow and little patches of green began to develop around the edge. The seeds, eggs, and aquatic insects in the muck began to proliferate and the pond was soon teeming with life; water striders, diving beetles, insects that make their home in rotten twigs, small red water spiders and a host of other creatures began the colonization process. Algae found the sun-warmed water to their liking and I had several outbreaks of algal blooms. The pond became, according to my family, a slimy stinking and unsavory mess. Not the beautiful crystal-clear water they had dreamed of. I relented and poured a bottle of bleach along with a few crystals of copper sulphate into it. The algae died, turned white and dropped to the bottom of the pond and began to form a decaying organic mantle over the stones. Dust also blew into the pond adding to the bottom mineral content. Winter came and I ensured that the pond was full of water before it froze solid in October. I waited expectantly until the next April when the ice would melt and I could see what had survived.

Springtime came, the ice thawed and the pond picked up where it had left off. The plants survived, greened up and expanded their territory rapidly as they sent rhizomes out toward the uncolonized centre. The insects sprang to life and set about their business of reproduction and survival. They expanded their dominion and complex predator/prey relationships began to evolve – the hunters and the hunted. The hunting and survival strategies were brilliant and varied and gave me hours of pleasure simply watching the dances of life. The water striders would stalk their tiny prey by darting toward it a few inches at a time then stop motionless on the surface until the target felt safe again. Then another dart forward until finally the prey was within striking distance. Territories began to develop with tussling and jostling taking place for hunting and breeding rights in different areas of the pond. Real estate is important to most animals.

Early on during the second year I noticed a few small snails had appeared. Initially there were so few large enough to see that I used to try and count them. It wasn't long before their numbers had increased and one could hardly find a few square inches that careful study did not reveal a snail or two. Snails are prolific. They seemed to thrive on the algae and other things that grew on the pond bottom and on the leaves of the ever-expanding higher plants. While algal blooms still occurred these were much less frequent during the second year and I only had to resort to the bleach bottle a couple of times. By the end of the second year the margins of the pond had been mostly colonized by the expanding higher plant territory and even the centre of the pond had a few shoots of green. A few new plants began to appear, presumably from seeds present in the mud; *Hippuris, Ranunculus, Eleocharis*. The pond bottom was now covered with dead and decaying matter and enriched with more wind blown dust and bits of grass spewed out by the lawn mower. I got fewer complaints from my family as the pond greened-up. Our neighbor's children loved it and used to come over with their bottles and nets to collect 'stuff' for school projects.

By the end of the third season the pond was a rich green, recognizable plant communities were beginning to differentiate; patches of dominantly *Carex, Eleocharis, Scirpus* and *Equisetum* along with associated specific insects that found one community type and its environment more or less to their liking. Territorial boundaries changed and the general free-for-all for space became less random and a bit more orderly. Floating aquatic mosses appeared and formed mats on the water surface. I occasionally had to resort to a rake to draw off great clumps of moss or the water would not have been visible. Much debris from dead plants covered the bottom and the stones were no longer identifiable. Algal blooms ceased and the wild eutrophic swings diminished. The pond was now fundamentally orderly and stable.

Since the third year there have been no wild algal outbreaks, the bleach bottle has been put away, the vascular plants have expanded, the mosses are easily controllable if they begin to trouble me and each year during spring and fall migrations, ducks drop out of the sky for a brief visit to feed on some of the ponds bounties – on a couple of rare occasions even Canada Geese have found it worthy of a visit. At the moment my principal task is to decide whether to arrest the natural progression of ecological change and intervene to ensure that a pond full of native species is still part of my garden. The pond is as natural as any pond in northwest Calgary. To claim otherwise would be arrogant. I dug the hole for the pond, (instead of glacial ablation) to collect and hold water, shoveled in a bit of muck and natural processes established a beautiful wetland.

If I choose to continue to have a wetland in my back yard then I must interrupt the natural process of change and arrest them by putting resources of time, energy or money into the undertaking. Such is the lot of humans as they manage ecosystems to ensure the ecosystems produce what they want. Agriculture and increased production of desired agricultural products is only possible through the input of labor and energy to prevent natural processes going in the direction that natural processes will. Ask any farmer who still works the land; he or she will tell you about their struggles with nature.

That is the nub of resource management: deciding what one wants from a piece of land and then figuring out what interventions are necessary to achieve the desired goal. Generally, science and logic are the easiest elements to address in resource planning and management. The really tough part for people is deciding what they want and what the goal is. Goals are founded on values and beliefs and these must be clarified – a difficult task for many folk.

As an addendum to the pond story I note that during the fourth year a tiny northern chorus frog found itself making a home in the pond. These tiny amphibians have a voice totally out of all proportion to their size. It never found a mate and spent morning, noon and night croaking so loudly that our neighbors complained frequently, and in good humor, about the racket. Only three years later did they 'fess up' and inform me they had found the frog during an outing with their children and had slipped the frog, unbeknownst to me, into the pond themselves. It has been a great source of neighborly joking and thus the pond has provided a study in social ecology.

On Reconstructing Native Grasslands:
A Community Act

Concepts related to the reclamation of disturbed lands have gradually evolved over the years and continue to do so at an impressive rate. Recent public awareness of human induced changes and the desire to maintain "fundamental ecological processes" (a currently important phrase) on the landscape have spurred the development of reclamation science even more so. Governments are promulgating legislation requiring reclamation and decommissioning and community-based groups are active in demanding the preservation of complex native ecological communities. Such is the flavor of the times for in many areas wealth is widespread and with wealth and the consequent lack of want among the populace there is an outward looking, a desire to preserve and maintain natural beauty, a desire to preserve pieces of land in a form that was widespread prior to the breaking of the sod, laying of concrete and erection of edifices of human development; prior to the digging, ripping, clearing, tearing, scratching, pouring and banging commensurate with 'progress'.

The people in these enclaves of wealth have a vision of what they value. They have a vision of nice homes and a pleasant lifestyle with all the city amenities; urban transport, parks, symphonies, the theatre, outdoor sports, community activity programmes for their children, good schools, shopping, safety and good neighborliness and pleasure. Safe drinking water, good health care, high quality waste disposal and adequate personal safety are rarely questioned; their presence is simply assumed. They are kind people, social conformists with enough edge to them to maintain their individuality and originality. They are not the human equivalent of the functional, though stripped down, Chevy model designed for the basics. They work hard, command more resources than they need and are not strident. With wealth and station comes a measure of public tranquility making stridency unnecessary; strings can be pulled in private settings, community goals achieved, and wants more or less satisfied. There is enough education and enough knowledge in the community, enough connections and enough unity of desire to achieve valued community ends relatively easily. There is a strong social conscience and social obligation; people volunteer in the schools, the churches, the community projects; they discuss, they communicate and they socialize. There is a commitment, a sense of community and a sense of what is perceived as the common good, the common weal. It is a civil society, as inclined to give as to receive, and there is a strong community spirit.

They also want safe parks in the immediate proximity of their home. They recognize that the development of their community and their personal land have come at the cost of eliminating previously undisturbed native ecosystems and they want these native communities incorporated into their parks so they can explore and enjoy the natural beauty unsullied by the human endeavor.

This is the social context in which I undertook a community based ecological restoration research project. These were the people I worked with, planned with, organized with and laughed with. I will offer another description about ecosystem change and adaptability within the context of achieving our goals of restoring native Rough Fescue grasslands in an area cleared for housing development.

Reclamation is really quite a difficult topic in some aspects and quite simple in others. Though some might claim it a science, a discipline, I suspect it is better described as a collection of methods, concepts and techniques gleaned from a host of disciplines – not all of which are sciences – applied to achieving one's ends of altering a landscape to some specific goal. That goal might be curtailing erosion, increasing slope stability, making disturbed land agriculturally productive, turning nasty deserted gravel pits into lakes or parks, detoxifying chemically polluted sites or re-establishing native ecosystems. Reclamation can be divided into three pursuits as defined by the National Academy of Sciences in 1974: *reclamation* itself involves the re-establishment of the productivity of a site; *rehabilitation* implies the land will be returned to an agreed upon standard in accord to a prior land-use plan; and *restoration* involves restoring a site to its original conditions.

It is not a large stretch of the imagination to realize that restoration is by far the most difficult of the three. Restorationists must not only fix up a disturbed site so it is stable and productive but they must put the same species back on the site and make it as it was before; this is no easy task given the complexity of most native ecosystems. In fact there is not a lot of knowledge about how humans can restore natural systems. There is no manual to go to in order to find out how to restore a given community. The literature is evolving and techniques are being developed but there is much confusion as to how to judge successful restoration and questions abound. Rough fescue grasslands fit into the category of complex systems that evolve and change over decades as a result of the adjustments of the biological component to changing physical characteristics of climate, soils, slope and topography. It is not one easily definable ecosystem that can be characterized, classified, established and judged. Rather the grasslands are a continuum of nature with infinite numbers of forms and mixes of species and abundances. Consequently the restorationist faces the first problem of deciding what type of rough fescue grassland is their goal if they want to be able to measure the success or failure of their efforts.

During my more reflective moments it occurs to me that the notion and image of humans restoring nature is enormously egotistical. What are we? Gods, who with a wave of our hand command the Earth to be what we wish? By changing a few things we are able to change the sequence and timing of nature's self-restoration.

The rough fescue grassland zone covers about 1400 square kilometers of southwestern Alberta where it occurs in the plains and foothills of the Rocky Mountains. There are numerous plant community types comprising the vegetation zone commensurate with the nearly infinite physical variation in slopes, aspects, soil, climate, nutrients, and moisture and wind conditions. These also vary over time as a result of interactions of the biological component of the system and the various disturbance regimes such as fire and grazing. It is complex and simply and honestly put, humans know very little about the inner working of these systems. The restorationist of rough fescue grasslands not only faces the challenge of infinitely variable restoration targets but must also deal with immense human ignorance to boot. The accuracy of sophisticated and complex interactive and predictive ecosystem models is shaky at best and typically inaccurate. Characteristically the more precise these models of natural ecosystems attempt to be; the less useful they are in general application. Classifications attempt to provide snapshots of the continuum of nature with its infinite variability, for the purpose of simplification to make them understandable. Plato addressed this issue over 2000 years ago in his 'Civitas', a tome well worth a visit.

Of all our native Alberta grassland types, the rough fescue is the most biological productive for forage. On average, under undisturbed conditions, it has been found to produce about 800 kg/ha of cattle feed. It makes wonderful ranch pasture, produces lots of tasty red meat and is truly beautiful to look at as the plants bend and wave at the beck and call of our frequent Chinook winds; catabatic winds, that come roaring over the Rocky mountains during the fall, winter and spring seasons as the cold dense western air crests the Rocky Mountains and begins to tumble down the front ranges to the prairies. Frictional forces warm it up and the biological systems in the area have adapted to rapid temperature changes and high winds. Sometimes the wind speeds reach well over

120 km/hour and temperature changes can exceed 40C in a particular day. Chinook is an aboriginal word meaning "snow eater" and when these winds come, people get a break from the intense winter cold and the animals get access to forage as the snow melts and reveals bodily sustenance for them. Not to be too anthropomorphic, but the plants also have to figure out when to start to break their winter dormancy because if they choose a period during a lengthy winter chinook they may well be faced with another extended -30C cold snap that will deplete their limited energy resources and finish them off. Life is tough in the rough fescue zone and only the hardy have survived; the unsuitable have been eliminated over the eons.

Humans have found that in spite of the high forage productivity of these native grasslands, the forage productivity of the land can be even further increased; doubled or perhaps even tripled, if the Fescue is removed and high yielding forage species planted instead. In the quest for increased yield many of these grasslands have disappeared and been replaced. Some estimates are as high as 90% loss. In addition to agricultural interest losses, urban development has taken its toll. Calgary is built in the Rough Fescue grassland zone and as a modern, rapidly expanding city, it consumes much native grassland.

This native grassland was the object of desire of the community I worked with. The community has a huge amount of 'green space'; about 40%, the highest proportion in Calgary I am told, and much of that green space is parkland and slopes too steep for development. These slopes are clothed in native rough fescue grasslands which are highly valued; however, the community wanted more native plants and was prepared to put the effort into meeting their wants. They did not know how to get them met. That is where I came onto the scene as I am also a resident, a new one at the time, of this community. How could I not get involved in such a project? As a plant ecologist it piqued my academic curiosity; as a professional biologist involved in applied ecosystem science I could see the possible conservation and reclamation applications at a broader level; and as a resident of the community there were social benefits. As an ecologist given to research and publication, the idea of having a research project right beside my home in an area we walked by during many evening strolls was more than could be refused. It has been said that people maximize their self-interest in the best way they know how to; this was a personal deal to beat all deals. To boot, the people were pleasant and enthusiastic to work with and vigorous in their efforts. We soon decided that the restoration site would be on steep slopes beside the elementary school where the children would be able to see and learn from our experiment.

I was left with the task of figuring out what to do and how to do it. A quick look around the expanding community revealed areas of rough fescue native grassland that were falling under the fate of the earth scraper as new parts of the subdivision were being prepared for construction. I thought that if we harvested the sod before the scraper got to it we could simply lay the native sod on the target site, the task would be complete and our goals met. I was, however, not sure it would work because the technique had not been tried before. With the help of the developer I was given the use of a belly scraper to try and cut the sod; the method failed as the sod got churned up and muddled as it entered. I then decided that I might try a standard small-scale sod cutter that can be rented at most equipment rental shops in our area. It worked and I got ribbons of sod that were 2.5 inches thick and about 18" wide. The perfect size for physical handling by a community volunteer group. The only problem was that I didn't know if the sod would survive a transplant or winter and if so, what species would survive. A perfect research project; lots of unknowns, an experimental technique, lots of hypotheses and no evidence at hand; simply theory!

I take you back now to my personal garden. I had planted a native aspen copse on one of the hills in my back yard but had not yet figured out how to get native understory – the plants that grow beneath the tree canopy. Serendipity came together and I decided that rough fescue grassland plants would be entirely ecologically appropriate as an under story. There are many aspen copses in the Calgary area and some consider these to be a separate vegetation zone. The aspen copse in my back yard became the research trial site to answer the questions of species survival, composition, and winter endurance before translating the whole exercise into a community project of grand proportions with no known probability of success. We rented a sod cutter, cut sufficient sod for my copse and laid the sod. If elegant experimental design is characterized by its simplicity, then I lay claim to the simplicity part of it. The elegance part is highly questionable. Time would tell whether the experiment was successful; patience is a virtue is such situations!

Spring came and I watched. My research project was now in my back yard and I could easily monitor it several times daily. As the snow melted and the sun warmed the soil, the sod gradually sprang to life; the grasses greened up, the more showy flowers came into leaf and bud and there was a general sense that success was highly likely, if not inevitable. The odds were in our favor and we decided to proceed with making plans for the larger re-sodding project. Nature is tough!

The restoration event was visualized as a community spirit building exercise in the tradition of the early community barn raisings when Europeans settled Alberta at the turn of the century. Ads were put in the local community paper seeking community volunteers with stout hearts, rakes, shovels, hoses, trucks and other useful implements. The City was contacted and they offered staff and equipment for site preparation; the parks department had in addition to their curiosity a powerful self-interest for if it worked and the process made economically viable for large areas, there would be a considerable reduction in maintenance costs and a reduced requirement for herbicide applications throughout the city. The developer offered more resources, the equipment rental company offered equipment and some of the construction contractors offered staff and equipment. Meat packers offered hamburgers; breweries, beer; pop companies, soft drinks; and individuals, other food. Provincial and municipal politicians were invited to attend our community event and were told they would be given a platform and a shovel.

The appointed Saturday was beautiful and sunny, perhaps a bit too hot for optimal sod transplants, and by 7 am people began to show up. By 9 am it was clear we would have lots of labor and the event would not fall flat. Young families with children, older couples, individuals, dogs, the occasional cat and the whole spectrum of community folk showed up; we had an army some 150-200 people strong. The sod cutting team got to work mowing the climax native fescue sod first so a greater depth of soil and more roots would come along as the sod cutters had a maximum depth of 2.5 inches. The sod transport team rolled, loaded and drove the sod from the donor to the recipient site as fast as possible to prevent drying and the sod laying team set to their task of cloaking the donor site with as tight butt joints as possible between the individual strips. Others staked the sod to prevent it from slipping down the steep slope. Another team was given the task of going to the donor site and digging up shrubs, trees and a few of the rarer, more individual and occasional species and inter-planting these on a sporadic basis throughout the sod to ensure that a more comprehensive representation of the original species was relocated. Hoses were rigged and the sod watered to increase the odds of transplant success. A marquee tent was raised, barbecues set up and refreshments and nourishment put in place. An air of conviviality combined with singleness of purpose prevailed; new friendships were made and old ones renewed. From a social standpoint the event was a winner and from a restoration labor perspective our task was achieved in record time. By the end of the day all but a small piece of the one and a half acre recipient site had been re-sodded and folk left with a sense of achievement and in high spirits. The next day was rainy and the completion effort delayed until a June weekend. That weekend a few hardy souls showed up and the sodding was completed in a couple of hours.

The above events took place in 1991 and I have had nearly thirteen years to monitor the experiment and watch what happened on our community project site; fourteen years at my private garden site. I offer the following summary of my observations as they are instructive in the area of ecosystem change and adjustment.

To begin with, I will note a few properties related to climax rough fescue grasslands and describe some differences between the donor site and the two recipient sites. The donor site was a climax fescue community on a relatively flat area. The soils were deep and the fescue formed large dense tussocks with few spaces of exposed soil between them. These few spaces were covered with a dense thatch from previous year's fescue growth and prevented the growth of many characteristic species common in earlier successional stages of the plant community. The community recipient site was on a steep southwestern aspect and it had received a thin topsoil dressing over the native till before the sodding took place. It had also received an herbicide treatment earlier to eliminate weeds. My private back yard recipient site was a small hill with young aspen trees providing some shade in areas. The topsoil was rich and deep; two to four feet thick. The main slope was gentle and the aspect primarily north facing. These site differences are important in explaining what changes took place.

Back to the community site. Very quickly – a few days – after transplant there was a check in growth, a slowing down, as anyone who has either laid sod or taken a cutting of a friend's plant and started a new plant knows. In the process of cutting the sod at the donor site many of the roots had been trimmed and the balance between the water and nutrient gathering organs and the transpiring organs altered. The roots could not gather sufficient water to supply the leaves with their needs and because they were only two inches deep they had little access to deeper soil water. Some of the leaves, particularly on the larger healthy plants and those species that develop deep roots suffered the most. Nature's way is to try and re-establish that balance. Some leaves wilted, withered, dried and died. Also not surprisingly it was the tall healthy leaves that suffered the most as they were exposed to the wind and had the greatest call on water resources. It was not long, a week or so, before close examination showed that the sod was beginning to 'catch' and the plants that had died back started to produce new growth. Slowly at first and then more rapidly as the roots began to grow downward into the soil beneath the sod.

The balance between the root and the shoot of plants is a wonderful thing and has been the source of much scientific curiosity. Generally plants growing in dry areas have a large root to shoot ratio while those in wetter climates have a much lower one. Many years ago a study showed that a single plant of Little Blue-stem (*Bouteloua gracilis*), a short grass common in the dry Alberta prairies, had a total root length exceeding 214 miles while the above ground part was only a few inches high. Other species commonly have root lengths that are many miles long. To make this more believable I will note that the length included all root segments including root hairs. A tedious calculation task this but rewarding work nonetheless as it places the brilliance and the balance of nature firmly in ones mind. Plants, like most organisms, do not expend more energy than necessary but they neither expend less than they must in order to survive. A good example of this can be drawn from gardening and lawn maintenance. There is a tendency for many folk with gardens to water their lawns and gardens frequently because they think the plants will grow better and more robustly. Nothing could be further from the truth for when plants have easy access to water at the surface of the soil they don't bother sending down deep roots into the permanently moist soil and they never develop drought resistance. When a dry spell comes they must still be watered frequently or they die or go into dormancy. This has a high water resource cost, as any city parks or waterworks department will tell you. With a bit more knowledge on the part of lawn owners, cities would have to provide considerably less water to households. A few good garden soakings on an intermittent basis are all that most gardens need. Nature is not resplendent with hoses to take care of native plants. If they can't get by they die.

After a few weeks I noticed some new species growing in the cracks formed by the butt joints where the strips of sod had been laid side by side and end to end. I watched these develop and soon the sod looked, from a distance, to be covered with weeds a foot or two tall with mustard yellow flowers. This development upset one of the neighboring families as they had expected their view to be one of waving native prairie covered with glorious wildflowers. Reality and expectations don't always match. They complained to the city and the city in turn brought this development to my attention. I had noticed this weed development earlier and was not concerned. I informed the authorities that these were simply common early successional weeds of the mustard family and they would be eliminated, perhaps in a year or so as the native sod grew horizontally and filled in the cracks of the butt joints with new roots, dead leaves and the wind from the neighboring construction area deposited dust. Rains, I said, would also help the process. Regardless, I decided it would be prudent to expend some time pulling these foreigners out in order to keep the peace. The task was easy and quickly executed. Closer examination showed that it was not just the mustards that invaded the cracks but a host of other common non-native local weeds; foxtail barley (*Hordeum jubatum*), couch grass (*Agropyron repens*), burning bush (*Kochia scoparia*), Lambs quarters (*Chenopodium alba*), Russian thistle (*Salsola kali*) all had a liking of the butt joint cracks where little else grew, resources were available and competition minimal. Few of these weedy plants gave me any concern because they were confined to the areas of the joints and showed no inclination to penetrate, grow and spread into the newly laid native sod. There was, however one exception: couch grass. Couch grass is an aggressive introduced grass that is not easily controlled. We had noted its presence on the recipient site before the planting and the City had kindly treated the site with herbicides to control weeds before sodding. Perhaps the spray had been unevenly applied but over the summer the couch grass flourished and formed three large patches in our newly sodded area.

I have watched the development of weeds in the cracks over the years and as expected it was not long before all of the annual weeds were out-competed and eliminated from the restoration site. They thrive only on disturbed areas in our region, waste places, but they cannot stand competition from the tough local perennial species. The Couch grass patches remain to this day having neither expanded their territory nor allowed the ingress of other species from the native sod. We had tried to eliminate the couch grass with herbicides a few years into the experiment but to no avail. The grass was killed in places but sufficient remained to quickly reclaim the former territory. We will see what happens over the next ten years. Will the couch grass succumb to the advances of the native plants or will the native plants be invaded by the couch grass? At the moment a territorial truce (or perhaps a standoff) seems to be the current bargain.

Other species have colonized the area over the years and some of these are also Eurasian introductions. Many years ago, over a century, early European settlers brought seed out to the new land to plant their fields; Crested Wheat grass, Smooth Brome, Timothy. These have some particular properties that make them attractive in pastures. They are productive, hardy, quite aggressive and easy to establish and grow. They are also highly prized species by many city parks and reclamation departments and by cattle producers for these reasons, and highly despised species by most folk who want only 'pure local native species.' Many environmental organizations, and some scientists, lecture and canvas hard and long on the evils of these creatures and of recent there has been a move to eliminate them from seed mixes. They are often claimed as aggressive weeds that destroy the 'integrity' of nature. While I admit that I would prefer not to have them as the part of our Rough Fescue prairie site it only takes a moment's reflection on the classification 'weed' to realize that a weed is little more than a plant growing in a place that humans don't want it to grow. Nature really does not care and even weeds were native in some location at some point in history. As for the 'integrity' of nature – what a funny notion! I can not imagine nature without integrity – or with it for that matter! Integrity seems to me to be a human value-based concept and I can't quite get my head around the idea held by some folk that nature should exist or change in accord with what humans think, want or wish nature to be.

The community restoration site, as noted earlier, is on a steep southwestern aspect. The top of the slope is bounded by a school playing field and the bottom by a cement sidewalk. Over the years a few of these Eurasian introductions have found a place in the fescue sod. They are not abundant and they do little to change the general character of the site. A few are interspersed here and there near the boundaries and at the moment they seem to have found a place in the current balance. Where the seeds came from I can't say; a passing bicycle, wind dispersion, mud coated boots of residents during an evening stroll in wet weather, passing animals. Perhaps they even passed through the gut of some creature. Regardless, to be a purist and try to eliminate them would create more angst than necessary. After all, if they got established once in the past it seems clear there will be similar pathways for them to get re-established down the road. I remain curious about their future and in the absence of a body that has run its course, I may be able to monitor it for many years to come and get farther into the story and see how the plot develops. The story of ecosystem change does not have a final chapter; at least it has not for several billion years in the past. For that matter it is not even clear there is or ever will be a plot to nature although it seems likely (given evidence from astronomy) that ecosystem change will at some point cease on our planet.

Initially the species composition of both restoration sites was similar to that of the donor site; lots of Rough Fescue and less abundant quantities of other grasses and herbs. Things looked promising for we humans, and our aspirations but that wasn't to be the case for long. Nature has a certain direction about it and that direction is not always in conformity with the direction that humans want or expect. Gradually species that were not present in the inventory of the donor site appeared. Happily they were mostly native species that form part of some native rough fescue prairies types in some areas. Some fairly rare and intermittent species in the donor site began to appear; anemones, shooting stars, *Thalictrum*, other grass species. The flowering species diversity of the reconstructed communities increased quite substantially from numbers in the 40s to numbers in the 70s. Where these species had come from one can only surmise. They may have been lurking in the soil of the donor site unable to compete against the vigor of the climax fescue or they may have been transported into the site by other seed dispersal vectors. Regardless, our transplant actions had changed the conditions and they found our disturbance actions to their liking and began to capitalize on the opportunities that our altered conditions afforded. We humans had increased, through our actions, the species diversity just as non-human nature might do with fires, landslides, erosion, animal burrows or grazing animals.

The relative characteristics of the donor and the two recipient sites undoubtedly had a large role to play in the changes in species composition and abundance. The donor site, as noted earlier, was relatively flat and uniform while the two recipient sites were not. The tops of the slope at the community site were drier than the middle while the slope bottom was much wetter because water always flows downhill and in the process of its movement it dissolves many chemicals and translocates them from one area to another. Seepage and groundwater movement are important in physical and chemical site differentiation and species in the biological world find some sites more agreeable than other. Fortunately not all species like the same thing so highly differentiated sites tend to host a higher diversity of species than uniform sites. The slope began to segregate into initially different abundances of the various species and over time to differentiate into distinctly different groupings of plants. Some species were only found at the top, others at the bottom. Not only did the species vary but also those species that grew across the entire slope showed variation in their growth rate and productivity. Where more space, light, water and nutrient resources were available the plants did better. Now ten years into the experiment there are some areas, mostly at the slope base, covered with lush fescue tussocks similar to the original donor site. Their species diversity is diminishing yearly as the fescue gains dominance. But different species dominate in different areas of the slope. Time will tell how powerful a force the fescue can exert over the whole site.

As in all matters of life one can get too much of a good thing. Where there are too many resources for a particular species' liking , the species does badly or dies and where there are too few resources the outcome is the same. A little bit like the children's story of the three bears and their porridge; some conditions are just right. In ecology we know these phenomena of 'too much', 'too little' and 'acceptable' as Range of Tolerance, Leibig's Law of the Minimum (when there are not enough resources) and Shelford's Law of the Maximum (when there are too many resources).

The Greek philosopher, Theophrastus, knew this natural differentiation process and made an exhaustive list of plant responses to the different environments long before Leibig and Shelford came along and stated it in the tidy words of laws. The knowledge probably predates even Theophrastus by at least several millennia to the beginnings of the Neolithic period and the advent of agriculture. Likely even in the early dawns of human development when hunters and gathers noted that some food types were more or less abundant in one area or another. The laws apply to all life large or small, animal, plant, moneran or protist. They are fundamental laws for humans as we try and produce food, fiber, aesthetically pleasing gardens, control diseases, wage wars, reclaim the areas we disturb or colonize environments in which the naked human body could not survive. These laws are the fundamental drivers of human technological development as we attempt to alter unsuitable conditions to make them at least minimally acceptable if not optimal: weaving, clothing, houses, saws, screwdrivers, shovels, fields, furnaces, fire, higher order machines, computers or nuclear energy. The absence of fulfilled needs or wants results in technological developments to find solutions to problems real or perceived. In stressful environments where the living isn't easy, organisms have to be resourceful to survive.

Life can get too comfortable and when it gets too much so, complacency sets in and individuals become less adaptive. The evolutionary record is littered with species that were unable to adapt to Earth's changes. Nature seems to provide lifetime warranties and best before dates but there are no manuals defining how long a lifetime is and no date stamps on nature's products.

Back to the fescue recipient site story. The recipient site in my back garden was quite different you will recall from the larger community site. My private one had thick topsoil and a bit of shade. The thick topsoil was able to hold more moisture and was less inclined to experience droughty conditions than the other site. Over the intervening fourteen years the site has not come to resemble the original donor site. The grasses are tall, thick and prolific and the majority of them are not Rough Fescue. *Elymus, Bromus, Poa* and *Agropyron* abound. This change took place in the first few years, the tussocks of Rough fescue diminished and one had to look closely for it. The taller grasses shaded fescue out with their luxuriant growth. Now, ever so slowly the rough fescue is increasing and spindly tussocks are in the early stages of development. There is an inevitable battle for resources going on and what species will win the war is not yet clear. The rough fescue certainly does not like the shady areas under the trees. Wild rose, uncommon in the donor site, has sprung up here and there and is increasing in abundance. *Thalictrum* is abundant in the shade and in areas where the sun strikes there is an abundance of native flowers: *Gaillardia, Cerastium, Agoseris, Erigeron, Dodocatheon, Sisyrinchium, Smilacina, Thermopsis.* Initially Dandelion and Thistle found the site to their liking but over the years I have had to spend less effort in weed-pulling exercises. The system regressed and changed to an earlier successional stage and now it is experiencing fewer wild compositional swings and is stabilizing. A different plant community from the original donor site is developing.

Such are the changes of nature. One is not better in an ecological sense than another; simply different. One may conform to what humans value at a particular point in time and another may offend. When one species loses out, yet another gains at its expense. When rough fescue expands its territory and changes the environment, many of the smaller herbs lose; when fescue gets disrupted and upset, others gain.

On Perfection, Imperfection, Diversity and Stability

Four little human concepts. Ones that generate a great deal of economic activity and a lot of complexes in humans. They cause us great problems, we humans. The psychologists love them and many an analytical couch has been worn out and many a throat talked dry trying to deal with them; imperfect partners, perfect bodies, perfection complexes. Engineers seek the perfectly balanced machines that last and gardeners, some of them, seek the perfect flower. The perfect diamond, emerald or ruby is a much-vaunted object and one that we humans, given the means, are prepared to pay dearly to own. Surfers seek the perfect wave and economists look for the perfect market.

Perfection implies success, orderliness, high quality, durability, stability, cleanliness, desirability, freedom from error, and predictability. Imperfection generally implies the opposites of these as humans attempt to place value on things. So important to humans are the notions of perfection and imperfection that they merit a short exploratory discourse in the context of natural systems, change, diversity and adaptability.

Let's begin in the context of chemistry. Humans have defined some one hundred and three elements comprised of atoms that we generally consider the building blocks of nature (even though we humans have artificially formed some of these and they only last momentarily). They are, most at least, quite predictable and constant in their properties. We know them to be comprised of neutrons, protons and electrons and we generally think of the atomic nuclei as durable atomic parts although sub-atomic particle physicists have played about and rather smashed our concepts as they measure transitory artifacts or effects that suggest atoms are not as stable and predictable as they first appear.

It is the electrons I want to focus on because to be perfectly stable each element must have a number of electrons in such perfect balance that they counteract the charge in the nucleus; negatives counteracting positives to generate stability. Not so in nature. Each element, except for the noble gases, is out of balance in this arena. Some elements have more electrons whizzing about their K, L, M ... N orbits than they need and yet others have fewer electrons than they need. Some have a positive charge and yet others have a negative charge. They are imperfect, inherently unstable, and not durable. Were they perfect they would be in perfect balance and stable, the charges would match. Many years ago, Dr. Heisenberg threw the imperfection situation into a bit of a tailspin when he noted that the electrons were not even orderly in their orbit distributions; he came up with the uncertainty principle and from that we got quantum chemistry, physics and mechanics.

Fortunately for us humans, and all life on earth, the chemical elements are imperfect. Due to this imbalance, this imperfection, one element is able to combine with another to form compounds of unimaginable diversity; acids, bases, salts, polymers, ring compounds, oxides, radioactive compounds, gases, liquids, solids. The various elements donate, receive or share electrons with each other and in their continual quest for stability they create an immense diversity of chemical forms each of which enjoys more or less stability in different worldly environments. Come to think of it, they indeed are responsible for the worldly environments for without these compounds there would be no water, earth or atmosphere. No wind, no waves, no rain and most certainly no life to worry about. No pain, no pleasure, no wealth, poverty, economics, wars, disease, want or emotions. No art, religion, ethics or science.

So relevant is imperfection to the creation of diversity that we should privately, or publicly if that is our inclination, sing the praises of chemical imperfection. Not that it really matters whether we do or not because that is simply the way it is on earth and we have no choice but to accept it.

Let's leave chemistry now and move on to the arena of life and see how the story of imperfection and diversity unfolds there. It is the reproductive process I want to focus on. Not any reproduction, only that part that results in the creation of two or more organisms from one. Somatic cells don't count as much in the bigger game of life.

The reproductive process is imperfect and prone to coding errors. The double helix is patently imperfect in its ability to self-replicate. The cytosine and guanine, thiamine and adenine do not pair up perfectly all the time in the right sequence and sometimes the chromosomes that they form do not always come out in the same number and form all the time. They're reasonably good at duplicating themselves but not perfect. Errors occur from time to time. The nucleic acid bases get muddled, out of sequence, broken or combined and each time that happens the resulting individual differs from its parent to a greater or lesser degree. Biological variation among individuals is the result and diversity of attributes develops in the population. Most of these errors are not particularly significant in the broader scheme of things but sometimes, just sometimes, they afford the individual some advantage in the particular environment where it finds itself located. Sometimes this individual, thanks to its reproductive errors, gets a particular advantage over other similar individuals. It is more fit, better able to capture the resources it needs, reproduce more efficiently, defend itself from predators or become a better predator itself. It is better able to get by in this less than kind world where strife and want are the rule and not the exception. Thanks to the error it is able to find abundance where want prevails. As Darwin and Wallace so aptly noted in their joint presentation to the Linnaean Society of London in the 1850s, evolution occurs and the advantageous traits of the individual get translated into the broader population of like individuals. The genetic error that resulted in a black colored moth in an otherwise white population gave a particular environmental advantage in soot-covered Britain during the Industrial Revolution. The black moth could not be seen on the sooty buildings and thus had a survival advantage over its highly visible and easily caught cohorts. The gene for "black" became prevalent and dominant.

In some species this advantage is transmitted through sexual reproduction and genetic recombination and sometimes, when the individual hasn't discovered the joy of sex, it gets transmitted through asexual reproduction.

Regardless, of the means of transmission, the earth because of its particular place in the solar system offers a lot of environmental variation to test reproductive products. The changing tilt of its axis, the uneven and limitless variation in the distribution of light, temperature, nutrients, water, humidity, elevation, moisture holding capacity, oxygen and carbon dioxide concentration and pressure afford nearly infinite variation in the range of environments and a near infinite playground of environmental diversity for these reproductive errors to test themselves and see if they can capture what they need to survive and reproduce.

Humans through their proclivity to simplify the world in order to comprehend it stick these individuals into hierarchical categories ranging from individuals to species, populations, genera, families, orders, communities, ecosystems etc.. None of these really exist in nature; they are simply abstractions of reality – order imposed by the human mind by our sense of similarity among individuals. We synthesize these individual observations into concepts, theories and laws. None of these are real but they certainly have helped us in understanding a bit about how the world functions and afforded us the ability to manipulate our environment through technology to suit our ends. Individuals are particular and more or less real while our higher classifications are abstractions and not concrete realities. As biologists we have happily lived with and worked with the fact that we really don't know what a species is in precise terms.

Our definitions are imperfect but they are good enough; they work – sort of. If you are interested in learning more about this, you might read Plato's *Civitas* or pick up some good books on epistemology, metaphysics, logic and the history of philosophy. Murray Gel-Mann, the 1969 Nobel Laureate in physics, has also written a wonderful book on simplicity and complexity entitled *The Quark and the Jaguar*. I highly recommend it.

We humans are interested in the accuracy, precision, predictability and usefulness of our concepts and we freely communicate these observations among each other thanks to our ability to take thirty or so nonsense sounds and combine them variously into words with meanings that we, by common consent, accept as the basis of verbal communication and language. Linguists will confirm that language is imprecise as evidenced by our frequent misunderstandings but it is good enough; not perfect but it works in our current environment.

Perfect systems do exist and perfection prevails in some mathematics but nature has tested the notion of perfection and found it wanting. Imperfection and inaccuracy prevail though out all aspects of the physical, chemical, biological and psychological world thank goodness. I'm not sure we could stand perfection; it is too static, non-adaptive, unchallenging, unchanging, predictable and ... boring.

I will offer two recommendations before we move on to the next section. Don't try my arguments in favor of imperfection with your significant other; they won't work. When you make a mistake be kind to yourself and don't judge yourself too harshly, errors result in diversity and some of these errors, happily, will open new avenues for the human endeavor.

On Ecology and the Academy

This field ecology and its connection to the human endeavor is not particularly valued in the halls of most biology departments at the moment because the current pursuit is much more directed toward bits of organisms rather than whole organisms and their relationship to each other and their environment. I lament this a bit in my more pensive moments for while academic biological research presently focuses on the sub-organism level, professional biologists in government agencies, industry and consulting firms increasingly have to play in the arena of the generalist as they go about the task of trying to understand and provide advice on the changes that humans do or might cause with particular development and management activity. I don't particularly concern myself with this void in biology because either the pendulum will swing back toward the stable centre of its ambit or the void will be noted and picked up by other departments. At the moment the latter seems to be the case as these field biologist organisms find themselves no longer fitting well in the biology department environment and instead have shifted their territory and are in an adaptive radiation phase; changing and finding new environments where they are more comfortable. This is the R and K selection of E.O. Wilson or the Biology of Small Islands problem of McCarthur in this Theatre in the Round of academic life. I still have good friends I work with in the biology department but I have found a happy environment among architects, engineers, chemists, other biologists, lawyers, planners, ergonomists, industrial designers, philosophers and planners. It is a stimulating and rich environment and we cooperate well within our interdisciplinary community in a kind of ecological dynamic equilibrium with our own controlling factors.

I consider myself a field ecologist and not a theoretical or a mathematical ecologist; a bit of a natural historian with a distinctly systems bias. I am a classifier of plants, animals, landscapes, climates, soils, stars or anything else in nature that intrigues me. Having classified them I am interested in how they work, how they are formed, what they do, how they all relate to each other and what makes their combined effort so changeable. From the time I was a small boy I roamed the hills and collected and pressed wildflowers, and speared and mounted insects. I used to hunt and fish a lot but I have put away the guns for going on 30 years now. I still fish because it gets me out among the mountains and lakes and affords the challenge of the chase. I like sailing because it takes me to beautiful and interesting places with good friends and demands that one be patient as they pursue their destination with only the wind as motive force. We have a powerful diesel engine that is usually silent but it is there, 'just in case'; much like the idea that a little knowledge is a limiting thing but it is better than none.

My particular academic approach may or may not have any peculiar merit or virtue above other approaches common in ecology, or for that matter other fields of enquiry, but it and its questions fascinate, please and intrigue me. I have found them useful both academically and professionally. The synecologist is a generalist with a little knowledge about a lot rather than a lot of knowledge about a little which is more commonly found at the other end of the spectrum of scientific ecology. A friend and colleague of mind refers to people of my ilk, in an oxymoronic sort of way, as 'specialized generalists' which rather pleases me for many linguistic twists contain an element of truth in them. He is of similar mind though he comes to his 'spot' from an engineering background and is interested in ecology while I have come to a similar 'spot' from a biology background; a biologist that is interested in how biology relates to the human endeavor which is largely what engineering is about. Synecology is a marvelously humbling pursuit because one soon discovers that one can never in a lifetime exceed the relative level of a babe in arms. Just as you think you have some natural sequence or phenomenon explained and modeled you are thrown a curve and, if you are honest, you are forced to go back to the drawing board. There is no philosopher's stone. Like the 'precautionary principle,' reason is a very good slave but a poor master – there are limits to knowledge and also different ways of knowing. All sources of good information need to be honored.

There is a certain likeness of human social and intellectual endeavors to ecosystems; just as we think we have it together the environment gets perturbed and we must change our views. It is a bit like the rough fescue transplant work we talked about earlier. The transplanted fescue community experienced a perturbation in its way of life, an upset, a stress, and it changed and readjusted to the new environment. Academic environments are similarly rich and given to intellectual perturbations that require a rethinking and an accommodation. Those that are dynamic and adaptive survive and those that are not fail. Too much intellectual tidiness and arrogance are non-adaptive in a changing environment. Dogma should always be questioned, students should always challenge professors and professors should always welcome thoughtful challenges from any quarter.

Robert Persig once defined the University as the Church of Reason and I rather like the definition because universities should not promulgate unreasoned ideologies; they exist to enquire, encourage and teach people how to think and reason free from political and other meddling. They should not teach people what to think. That is what society supports them for and expects them to do. There is a kind of dynamic tension between society and universities, an uncomfortable shuffling, but it has been quite some time since we have experienced a rout of universities in free and democratic societies. Either universities are skilled at marketing their merits or the patrons believe they are relevant and worthy of support. It doesn't particularly matter in the bigger scheme of things because universities will continue to exist whether or not the physical structures are desanctified by economic starvation. Like the churches, mosques, temples or synagogues of God, the real University Churches of Reason are comprised of the people and not the buildings.

Even the most intellectually repressive political regimes have had no more success in stopping the human endeavor than the great environmental disruptions – human caused or otherwise – have had in stopping the course of nature. On the contrary, radical change has resulted in brilliant new solutions to old environmental problems.

A little plant we know as *Rhynia* appeared some 380 million years ago in the tension zone between the Earth's water and the land and so began the colonization of Earth's land based environments. Single celled organism discovered there was some merit to cooperating and sponge like organisms appeared; eventually complex multi-cellular life found a place under the sun. Interestingly each of Earth's major mass species extinctions we know about has been followed by a golden age of biological innovation: coelacanths, cartilaginous fishes, bony fishes, reptiles, birds, mammals and humans; eukaryotic blue-green algae, nucleate algae, spore bearing mosses and ferns, sexual reproduction and genetic recombination, dicotyledonous flowering plants, monocots. One crafty orchid species is so specialized it mimics the female of a bee species and so attracts the male who in turn tries copulate with it. In this process of brilliant deceit the orchid gets pollinated and the bee gets frustrated. The dandelion, our common weed, found a superb genetic expression for successful competition in order to ensure the genetic formula housed in the egg cells does not get changed and lost by the fertilizing males. It has developed a wonderful scheme. The pollen grows down the pollen tube to the egg and stimulates the egg to double its own chromosome number – but not recombine with the male chromosomes and so mess up the genetic formula. Just to keep a bit of diversity and adaptability in the genetic code, a hedging of the bets on change, the male sometimes wins, and the pollen combines with the egg.

Were I a betting man, I would place much higher odds on the longevity of the dandelion species than I would on the orchid; it is less habitat-specialized, less genetically specialized, and more diverse in its form; more plastic, more resilient, more widespread and more adaptable. While specialist strategies have a brilliant day in stable environments, generalist strategies are much more advantageous and enduring over the long term on this changing planet of ours. Consider the lowly Cucarachas, the cockroaches, one of the most successful groups of organisms if durability is the measure. Interestingly, it is one of the creatures most reviled by humans.

Perhaps there is a lesson for we humans to learn from the orchid, the cockroach and the dandelion: don't put all your eggs in one basket; hedge your bets; keep your options open; buy your food in the grocery store but don't forget to plant a vegetable garden! These are the instructions that we humans gain from our curiosity, our enquiring minds and our ability to wonder. When we get good advice we should heed it!

On Environments

Man the classifier has, from time immemorial, attempted to classify his environment, to pass judgment on it and manipulate it to suit his needs. Artistic representation has schools; schools have periods based on form, color, development phase. There are different schools or approaches to philosophy, dance, music, sociology, politics, psychology, education, medicine, mathematics, and so the list goes on. Furthermore, each of these groupings and each of the sub-areas have their proponents and detractors. More often than not, each school formed in some environment and that environment affected its development, be it political, social, intellectual, climatic, terrain, biological or some combination of them all.

Language is greatly affected by the range of environments speakers find themselves in, and the words and fineness of the classifications reflect the importance of particular aspects of their environment. The more important a phenomenon or object in the every day lives of a people, the more closely they observe it and the more words they develop to describe the detail of the situation. There are a lot of words to describe a lot of things and phenomena. Since its development in humans, language always has been a busy and important human activity that has continually, and will continue to change. As people of different language groups come in contact, both languages are influenced and changed by each other; words are added, deleted, or altered. The concepts these words describe also evolve as the experiential world of the language groups become bigger and more diverse. When individuals of a language group are introduced into the environment of another language group these 'weeds' change the character of their host society and in turn are changed, and in turn, if the 'weed' returns to its original language group it brings change to the group of its origin. So it is with the culture of the people in the language group. The more they come in contact with diverse situations, the more they change. Experience broadens and the broader the experience the bigger the world and the bigger the world, the bigger the world-view. Big world-views generally generate more knowledge and experience that can be called upon when the society faces a problem in their environment. Ideas and experiences are powerful things in the game of human adaptability and the more people that subscribe to a language group, the more diverse and plastic that group becomes. Languages, and their cultures, that are plastic, changing and adaptable gain supremacy at the expense of those with few speakers

and narrower experiences. Any cultural anthropologist, linguist or ety-
mologist would confirm there is a global loss of small language groups
and cultures as they come in contact with larger ones and are assimilat-
ed. Changing unstable environments favor the diverse generalist group
that can find ways to adapt to the change.

This is the problem of the specialized orchid who is brilliantly suited
to a stable environment in the jungle versus the lowly weedy dandelion
that has found a home in most areas of the globe. I hope the people who
live near where the orchid grows become extremely wealthy so they will
value it enough to protect it. If they don't get wealthy shortly they will
change its habitat in their endless effort to scratch out a living and meet
their basic needs. I wouldn't rule out the possibility in the current human
environment that the needs of the local people will get pushed aside
and the needs of the orchid attended to first. The needs of the powerless
have always been predicated on the wants of the powerful. Power has its
privilege. In some quarters orchids are valued more than people. I certainly think it is worth trying to keep it around as long as possible
because I value its presence, I admire its ingenuity of expression to the
environment it found itself in. It is a thing of beauty and I might learn
something from it.

The future for highly specialized orchid does not look particularly
bright but the dandelion on the other hand ... check your back lawn, the
roadside, and the parking lot. I'm betting on the dandelion but I hope the
orchid survives.

Unlimited diversity in our wonderful world makes for infinite variability of environments. The rarity of the orchid and the abundance of the
dandelion illustrate the problem of species and how they adapt to the
diversity and changability of earth's environments. Consider how species
adapt and evolve to environments created by mountains and plains, hills
and depressions, and valleys and lakes. The unlimited arrays of rocks,
igneous, metamorphic and sedimentary, are arranged in vast groupings,
yielding a huge variety of geomorphic types which are constantly changing as they come into contact with the wind, the rain, the sun, water currents or ice. All forms and expressions of climate and weather; all agents
of change; acidic, basic or neutral soils; coarse, medium or fine textured
soils; salt, brackish or fresh water; polymictic, dimictic and monomictic
lakes that are large, small, shallow or deep in all areas of the globe in all
climates; large oceans, smaller oceans with larger and smaller islands
with lesser or greater terrain diversity on each; north or south facing
slopes that differ a lot the farther one moves north or south from the
equator, east or west facing slopes that differ according to the prevailing
climatic and weather patterns; intertidal areas that, depending on their

position on the beach slope and the time of the month and year experience greater or lesser exposure to the water or the air and the attendant local attributes of these.

Each of these environments has different degrees of light or shade, color, concentrations of gases, and albedos that absorb the heat of the sun at varying rates and become warmer or cooler as a result. Each has different concentrations of the hundred or so chemical elements and the millions of chemical compounds. Each environment has different energy levels and experiences variable forces and gravitational influences. Mountainous areas become flat, flat areas become valleys with rivers, or flat areas become mountainous as our landmasses and the Earth's tectonic plates collide with each other in their dance and struggle for stability. Some areas experience more and rapid change than others for brief moments in history but all areas on Earth change – a lot.

Before we leave this matter of diversity in the physical and chemical environments of the Earth, let's for a minute observe that I have offered a brief list, a classification of some of the various environments and their parameters available. It is a snippet of a classification and, as noted earlier, classifications are generalizations, simplifications of the particular; groupings based on perceived similarities and likeness of things. Each of these is, for all purposes, infinitely variable along a continuum of conditions. Even the sun's energy output is highly variable. There is no shortage of different environments to act and be acted on by the biological part of the ecological snapshot at any moment in time.

Mathematics, the perfectly constructed system of the human mind, tells us that the possible number of interactions among different interacting things is a function of mathematical powers. With each new parameter added, the possible number of options, the possible diversity of different environments, doubles. Diversity increases exponentially. It is like the crafty Indian sage who was consulted by a powerful ruler who faced a difficult situation. The sage, after solving the ruler's problem, so pleased the ruler that he was offered what ever he wished for a reward. The sage, knowing a bit of mathematics and having some familiarity with a common game board of 64 squares, wished to appear humble so he made the following request. Give me one grain of wheat on the first square, two on the second, four on the third, eight on the fourth and so on until all the squares are filled. The ruler, appreciating the humility of the sage and knowing little about mathematical powers, readily granted the request and ordered the wish to be filled. He soon discovered that he did not have enough wheat for reward; 1, 2, 4, 8, 16, 32, 64, 128, 256, 512, 1 024, 2 048, and that is only 12 squares . . . 4 096, 8 192, 16 384, 32 768, 65 536, 131 072, 262 144, 524 288, 1 048 576, 2 097 152, 4 194 304,

8 388 608, and that is only 24 squares. I won't, go on ... you get the idea. Play it out if it pleases you. If I remember the moral of the story correctly, while the king did not know a lot about mathematical powers he did know a lot about power and how to exercise and command it. The sage lost his head and the king kept the grain. Power has its privilege and knowledge has its risks.

Nearly an unlimited array of different environments can be found and they, to put a small, but general, bit of order on them, include: arctic, sub-arctic, temperate, sub-tropical and tropical environments in marine, aquatic or terrestrial contexts. Marine and lake environments include submergent and emergent contexts while terrestrial environments include xeric, sub-xeric, mesic, sub-mesic, hygric, sub-hygric expressions. Further expressions come about as a result of light and chemical variation.

It is a useful approach when one is trying to explain nature for fun, pleasure, profit or curiosity, to look at the following broad categories from both a physical and chemical perspective: climate, geology, soils, hydrology, and topography for all of these go into defining the water, light and nutrient requirements necessary for life. It is also useful to take a guide from the newshound and, in describing a situation, ask yourself the four of the five fundamental questions: What? When? Where? And Why? When it comes to describing and explaining systems which include life, organisms, you must add the final press reporter's question; Who? Only 'who' applies to living things and 'who', interestingly, applies both to groups or individuals. Natural change and the errors of the genetic code take place first at the level of the individual and only then do they find their way into the population and its higher levels of variations, sub-species and species.

On Species and Environments

As stated, there are a lot of different environments on this globe of ours; far more than there are species so each species gets a shot at occupying more or less of one or several environments. Some species, those that are particularly adaptable, variable and plastic occur in many but no one species gets it all and some – perhaps most – at least over long periods of time, end up with none. These we call extinct species; their genetic fire has been extinguished and they are cold! If the successful get too greedy they sow the seeds of their own demise. Their demise in turn makes more resources available for others to capture for indeed theoretically the Earth's resources are limited. In the senior game of life everything loses, or at least does not get it all. That is the way the rules are written.

Take what you need or find something useful that no other creature wants or can use. Create resources where there were none before, find a way to survive in a niche that nothing else occupies and find a way to capitalize on it and exclude others if they don't help you in your pursuits. If they are useful encourage them and work with them but don't let them get the upper hand or you lose.

That is the problem faced by our orchid. It, like all things, has its moment in history before the changes become too great for it to tolerate and it must either change or be eliminated. Unless one believes in uniform creation they must conclude that the predecessors of our orchid with its elegant adaptations developed a suitable response to changing environment and found a workable solution to their reproductive predicament. Reproduction is a high energy and tricky, messy business but without it the future of a species is bleak. It is a dirty job but at least some individuals of the population must do it.

If you are not interested in orchid reproduction then drop into the local bar, order yourself a beer (or the libation of your choice), sit down and watch the mating game. Watch the folk that are still players, still interested, not yet disenchanted, not too burdened with hurt, not too cynical and don't have too many hang-ups. Those that are still players for whatever their reasons; those that still want to step up to the batting plate and take their swing at the ball and see if the are still capable of a home-run tend to congregate in such areas as they believe, or hope will yield the results they want. Sometimes someone gets lucky. Bars are good places for these observations because you can be unobtrusive in a bar. No one questions why you are there. It is much more difficult to be unobtrusive when you crash a gathering of the community literary society. Besides, participating in the local book club requires more than just the ability to lift a glass of cool beer from the table to your mouth, give a pull and swallow.

In most polite societies you don't get to see much of the completion of the copulatory process but you get a good shot at learning what mating strategies are presently adaptive and non-adaptive; which hold promise and which are dead end non-starters. All players play and some strategies are generalist and some are specialist. Some are aggressive-aggressive, aggressive-passive, passive-aggressive, passive-passive. There is always a risk that you might even learn a trick or two that might just be helpful or beneficial at some later date. If you're sure you won't then you are out of the game and coasting on the back-slope. Thank God for dreams and memories of earlier days.

If an organism gets too specialized, or overly specialized, in similarly critical pursuits such as nutrient, light or water gathering or dispersal of its offspring, it runs the risk of a brief journey on Earth for nothing here is particularly stable for long. In the short-term fires, landslides, earthquakes, volcanoes, droughts or floods keep most areas in a constant state of disruption. With each disruption formerly captured resources are freed up and available for those with the ability to chase them. Disrupted biological, chemical and physical environments are hives of activity in the colonization and survival game. By and large, generalist organisms win initially in the chaos of disrupted environments and specialists do better where there is a bit more order, predictability and stability.

Let's revisit the importance of errors in the genetic code because in our earlier discussion on perfection and imperfection we saw that perfection was non-adaptive and imperfection adaptive. Perfection is non-plastic and non-adaptive. Errors result in diversity that is generally adaptive at the population level and perfection results in uniformity that is generally non-adaptive. Thanks to coding errors in the reproductive game individual variation occurs and, if advantageous, it stands a chance of getting transmitted into the population. Occasionally these errors are of such a magnitude that new species, or at least what humans call new species in their classification exercises, arise almost instantly; a change in the chromosome number, a radical change in the arrangement of the amino acid base pairs, the combining of chromosomes from different organisms that normally wouldn't mate together because they were genetically at the margins of compatibility.

The larger the number of errors or the greater the individual error, the lower the likelihood that it will offer the individual an adaptive advantage. Nature dispenses with disadvantageous errors quickly, is more or less indifferent to errors that are environmentally neutral and offer neither advantage nor disadvantage, and is quick to pick up on the happy beneficial errors. Remember Charles Darwin's description of the white moth species that found its home on the clean buildings of non-industrialized Britain (page 35, "On Perfection, Imperfection, Diversity and Stability").

I recently had the pleasure of visiting a pacific beach in Ecuador just south of Esmereldas with an Ecuadorian colleague and his family. As is my way, I got up early in the morning with my notebook and pen and went for a walk along the beach. At the upper reaches of the beach the sand was a dry, mottled light gray color. With some of my steps the sand briefly came alive with motion, something moved. I looked but my eyes couldn't discern any creature that might have been the source of the motion; everything appeared, even on close examination, to be nothing but undifferentiated mottled dry sand.

Now, neurologists and others that are specialists in human vision, tell us that the human brain, when confronted with an object, first of all scans and records edges and only after getting a snapshot of the object's general form does it focus in on the detail so I decided to sit down and observe a small patch of my mottled sand. It wasn't long before a brief motion occurred again and this time I kept my eyes on the organismic doughnut and not on the sandy whole. Having once fixed my object I scanned it in detail and was delighted to discover an absolutely motionless outline of a smallish crab so perfectly blended in with the sand that it was all but invisible except when it moved. Why it moved I don't know; perhaps it targeted some prey, perhaps it sensed local danger and decided to get out of harm's way. What ever its motive, it moved and when it moved it was visible to me and probably also to passing birds that might enjoy a crab dinner. When still, it was invisible and basically secure; when it moved it took a risk. During the time I watched, it moved very little. It had found a balance between risk and reward and thus far it had got the balance right, it hadn't become a dinner. What a perfect match this little creature was to the environment it found itself in. Its color pattern protected it quite well from the watchful eye of predators and also made it quite difficult, in conjunction with both speed and motionlessness, to be spotted by unwary prey. I watched it stalk and capture savory passers-by in the most elegant hunting strategy I had observed for a long time. Before continuing my walk I decided to check out the distribution of it and its colleagues along the beach slope. Where the sand changed color due to a higher moisture content that made it darker I could find no other similar individuals. Perhaps they had decided to stay in their holes until the sand dried to the right tone, perhaps the environment was no longer suitable for the survival of the crab and those individuals of the crab species that tried to colonize it got eliminated; perhaps its range of tolerance to the changed environment didn't make the lower beach acceptable for any one of numerous reasons. Instead of my little crab, a larger and darker type of crab occupied the lower beach. I didn't stay to study the larger species but instead continued my walk.

I hadn't walked far before I noticed a tortugeta, a baby sea turtle, slowly struggling in the early morning sunrise to make its way from the place of its hatching to the sea edge some 30 meters away before it was consumed for its protein content and flavor. Sea turtles are tasty, so I'm told. I sat down and again observed my tortuga's progress. The tortuga was dark and the upper beach it had to traverse was generally light in color although due to slight depressions and former wave ripples there was a bit of shadow from the rising sun and the depressions had not dried quite as much as the heights; they were a bit darker. The turtle was quite visible as it went about its business and I had not watched long when I noticed that my little fellow wasn't stupid. His motion wasn't random; he had a plan; probably innate, part of being a turtle. Having freshly emerged from his or her egg, life experience was not vast. Oceanward progress was slow but the tortuga would traverse the dry bright spots as quickly as possible and make its way toward the darker depressions where it would rest for a bit in the larger ones before continuing on. I don't know whether it targeted the depressions because it felt more secure there or because it was cooler or moister, but seek depressions it certainly did. So in favor of the small depressions was the 'tortugeta' that its path to the ocean was zigzag instead of direct. It would move horizontally across the beach to go to a depression rather than shoot the fall line and minimize the travel distance. Sometimes the shortest distance makes for the shortest life and the longest distance makes for the longest life. On his journey he had to cross a minefield of large crab burrows and occasionally one of these crabs would race out of its hole and attack the tortuga who promptly pulled its legs and head into its new shell. The crab, unable to do anything with the turtle in a timely manner before increasing its risk of becoming prey performed a lateral arabesque and hied itself back to its lair. The turtle, after a discrete waiting period, continued its seaward journey. Turtle shells offer an advantage to turtles. To complete the story, as the turtle got closer to the land-sea interface where the sand was wet from previous waves, some fortunate combination of a few smaller waves resulted in a larger, higher energy wave that reached farther up the beach, lifted the turtle and transported it into the ocean as the backwash and suction of the water followed its rules.

The ability to disperse one's genes and their errors for long distances and over wide geographic areas does not always result in either increased frequency or diversity of genetic expression in populations. A small change in a large population that offers a small advantage to the large population likely does not affect the overall population that much. If it does not get in the way of survival it may be retained by that population as part of the population's overall genetic diversity; part of its genetic memory. Perhaps in some future environment it may offer a greater advantage and 'be selected' (have a genetic advantage) and consequently become more widespread. That is part of plasticity and adaptability of a species. Similar scale changes would, in a small population or species with a restricted distribution, logically have a greater impact. A small fish doesn't play much of a role in a large pond but the same fish in a small pond is a force to be reckoned with.

Reproductive, geographic, temporal or social isolation also play a role in increasing variation. When individuals of a population find themselves separated from the larger group and unable to breed within the larger gene pool, the new population formed by this isolated population gradually differentiates and changes from the original population. The mistakes happen in the code and the new environment acts on them and selects those mistakes that better suit the new population to it. The differentiation may be small initially but over time it can become quite large, so large that if the two populations are brought together they will no longer interbreed and we have on our hands what we would call a new taxonomic unit. It might be at the variety, sub-species, species or perhaps even at the genus level. The larger the differences, the higher the taxonomic unit we would differentiate it at because that is how humans have arranged their ideas of hierarchical order on nature. Small but identifiable differences get classed as varieties of species, slightly larger differences at sub-species, bigger yet and if they can't interbreed without human intervention, we consider them different species. Bigger differences yet and we drop them into even more general units; genera, families orders, classes. As a young fellow I recall the mnemonic to help remember the top down classification system 'Kids play catch over farmer Green's shed'; Kingdom, Phylum, Class, Order, Family, Genus, Species.

So how do these individuals get separated from each other and the whole process of differentiation of populations begin? Chance? Design? The Earth's changing face? Definitely yes in all cases!

Let's take the case of submarine volcanic activity some distance, per-haps a large distance, from a huge landmass and its surrounding oceans that are well populated by organisms. The volcanic activity continues and gradually the lava pokes its head above the sea level and dry land forms where before there was simply ocean. Our new land mass gets exposed to the elements characteristic of land environments; rain wash-es over the lava and leaches minerals from it and occasionally collects loose particles of rock and washes them lower down the slope where they get trapped in slight irregularities in our terra nova. The volcanic activity continues and the island develops a volcanic cone that spews volcanic dust over the island or, at least, parts of it. Winds, whatever their prevailing direction, pick up the dust and deposit it in different areas and in varying depths according to the configuration of the terrain and its influences on wind patterns. The irregular surface of the terrain changes the wind patterns and forces and wind flow changes from laminar to tur-bulent. In the eddies of changing wind energy levels the Aeolian wind-transported material drops out and onto certain parts of the island. Over time the rains establish channels where water gathers and flows in its short journey to the sea. Some areas of our island remain wetter for longer periods than others. The sun warms the land and those areas that are dry experience a greater range of temperature variation than those areas that are wet because land heats up and cools down much faster than areas that have a lot of water in them; it takes a lot of energy to heat up water. Volcanic bombs get blown from the cone and larger rocks are strewn over parts of the island creating irregularities on the land surface with areas that are shaded from the sun's rays. These areas have lower light conditions and since they have less direct sunlight influencing them they remain cooler and moister. The exposed rocks act as water col-lection devices and the rains run across and down the impermeable rock and collect at its base and the protected bases dry out more slowly after the rains. The wind influences the bases of these rocks to a greater or lesser degree; some areas are wind protected and others more exposed. Tremendous physical and chemical habitat differentiation has taken place on our new island. The various aspects of the prevailing and local climate, geology, topography, and hydrology have all taken their toll on the exposed land and it has differentiated a lot from its original condi-tions.

Meanwhile at the oceans edge the sea swells have transmogrified into powerful waves as they approach the shallows. The waves have lashed the rocks breaking bits off, changing their shape, changing their chemical composition as the land takes a continual battering from the forces of water. Fierce storms bring bigger waves to the sea edge and the erosion forces reach higher up the island margins. Tides further differentiate things as the intertidal island edges are exposed to the air or submerged under the water for shorter or longer periods of time. Areas higher up the beach are exposed to the sun's rays for longer periods and have more fresh water washing over them during the rainy periods. They experience much greater temperature ranges due to their exposure to the air than those areas that are submerged or exposed for shorter periods and the light conditions are quite different in each of these areas. High energy and low energy environments differentiate, tide pools develop that heat up tremendously when exposed and cool down rapidly as the waves and tidal waters overtake them. Since the tides cycles don't quite correspond to the solar cycle some areas are exposed at different times of day on a quite regular basis. Our intertidal area has differentiated into a vast array of conditions and environments, according to the various forces acting on it.

The land and the ocean are also connected in other ways than just the intertidal zone. Crashing waves at the beach edge splash up on the land dousing it with more or less seawater and salts of calcium, sodium and magnesium and other chemicals. Winds passing over the ocean steal moisture from it and during storm events this water from evaporation gets enriched with salt water from cresting waves in the process of air sea transfer. The topographic shape, configuration of the land, determines which areas of our island are bathed in the moist air; some areas feel the full force of it and other areas lesser influence.

These are only a few of the physical and chemical influences that cause habitat differentiation in our newly formed island. Many more act and many more environments are formed. The submarine environments have also changed radically in our bit of ocean as the bathymetry was altered by the vulcanism. What used to be only the deep sea has become an immensely rich array of different terrestrial, aquatic and marine conditions; countless different habitats.

We haven't yet situated our island in any particular location on our globe; it might be in the arctic, the tropics or any one of the areas in between. It might be in an area influenced by warm or cold oceanic currents or counter currents; the Gulf stream, the Humbolt current or it might be influenced by major high or low pressure climate cells, and their associated winds; the Westerlies, the Northwest Trades. Any number of other natural responses to large or small bits of land plunked on a globe that is in the order of eighty percent water, the water's movement over the sphere could influence the land. Perhaps it is in the path of hurricanes or typhoons that transfer heat from the tropics to colder areas in more northern latitudes. It could be anywhere on the energy spectrum of Earth as nature tries to balance the heat energy and the resultant forces.

It doesn't really much matter where our island is situated because regardless of location, the general patterns and process that act on our island are more or less the same; they differ primarily in the degree of influence they play.

Our island is an area with a huge number of resources that have not yet, according to our rather simple story, been capitalized on and captured by living things. In reality, living things would have found a home on this island long before this degree of habitat differentiation had taken place. Nature and natural processes don't leave the land alone for that long. Let's leave our island alone for a while now and revisit biology.

Nature is profligate in her abundance, generous with her resources and not given to squandering opportunities. Life has found a way to capitalize on the Earth's physical, chemical and energy resources from the tops of the highest mountains to the depths of our deepest oceanic areas; from the highest latitudes of both poles to the equator and from the driest deserts to the wettest regions. Life is there. Nature's biological innovation is, as far as we know, without precedent; every manner and kind of innovation is utilized in the resource capturing game that all living things must play each moment. Alternately they become just a resource to be used by others still at the banquet table rather than remaining at the local table and being both a resource user as well as a resource. Even the air is filled with life and not just birds and insects; it is filled with microorganisms, seeds and spores carrying amino acid base pairs of unimaginably different combinations far and wide over the globe.

In ecology we speak about the biotic potential of an area; the sum total of all creatures, viable parts or creatures, their seeds, spores or fertilized eggs that can possibly reach an area. Bits of life that, if they land on a particular differentiated bit of Earth's diverse habitats, might have a shot at capturing the resources they need and setting about their business of duplicating their base pairs along with the occasional beneficial coding error.

The biotic potential of an area is immense. The winds, rivers, oceans and their currents, waves, glaciers, icebergs, gravity, erosional processes, landslides, avalanches, mud flows, and so on pick up and carry biological passengers to all areas of the globe while changing water levels make islands or join bits of land and increase the opportunity for forms of dispersal to change. Some of these passengers have adaptations that make them better suited to one form of transport or another. Light spores or seeds are picked up handily by wind and heavier seeds have elaborate adaptations to increase their surface area to weight ratio and make them aerodynamic enough to fly. Some float and get transported to new locations by water such as the cocoanut and the poplar seed. The physical opportunities and the biological adaptations that have developed to take advantage of these opportunities are both awe-inspiring and mind-boggling.

Not just physical processes contribute to the biotic potential of an area. Some organisms walk in, fly in, swim in or hitch rides on the backs, bellies, fur, feather, flesh, guts, blood or skin of other organisms. Birds and other animals eat seeds as they travel about and pass them through the gut; some survive to start in a new location. Yet other seeds have adapted to require passage through a gut before they become viable, they need the different environment offered by the gut to break their dormancy and make them viable. Microorganisms colonize the blood or mouthparts of their host and get transmitted during feeding exercises and organisms like those that cause malaria, yellow fever, dengue fever or rabies are spread. Marine and terrestrial mammals and animals and plants of all manner and kind traverse the seas, air and earth and each is a host, a guest, a friend or a foe to an entourage of other species; spreading them about, conveying them from one place to another. Some stick together for life, others are passed through the alimentary tract, the mouth, the sexual juices, and the blood. Others get picked up and eaten at the dinner table as they fall prey. Still others are rubbed, torn or pulled off because they are annoyances to their host or they simply let go because they like their new location better. Chance plays a role and sometimes, just sometimes, the new location is found agreeable. Probably more often than not it is disagreeable and the effects of the transport end there.

Nature has discovered that individual survival is not a sure thing. In fact, survival is at best tenuous and to cover the bases of chance most organisms are prolific in their production of offspring. Reproduction is a big, high energy business and most organisms invest huge amounts of energy in it; showy flowers to attract pollinators, elaborate feathers, fur, designs, patterns and colors to attract, guide, repel or encourage one sex or the other. Smells repel, protect, attract or deceive, and tastes and sounds of a similar array have similar purposes. There are sounds that can be heard by most organisms and others that are beyond the Hertz ranges of some; smells that are detectable or undetectable depending upon the sensitivity of receiving organs of each species. Sight, sound, smell, touch and taste; all the senses are capitalized on by life in a virtually unlimited array of transmitters and receivers to suit them to their environment and define the environment they are suited to.

Some plants produce hundreds of thousands or even millions of seeds, others produce few larger ones. Herbivorous animals produce more offspring than carnivorous ones and top carnivores produce the fewest of all as organisms are fitted into the feeding order of things; into complex food chains and food webs. There are first, second, third, and fourth order carnivores; herbivores that eat only one species of plant or only a part of it and yet others that are generalized feeders that are indiscriminate in their taste and will get their needs met from most organisms. Whether specialized feeders, generalized feeders, opportunistic feeders, selective feeders, or preferential feeders, all manner and form of hunting and gathering strategies are represented: stalkers, runners, trappers, netters, poisoners, biters, swallowers, and tool users.

The range of mating strategies and reproduction mechanisms and their use in plants and animals would make even the most liberal-minded human blush. The mechanisms are elaborate and varied. Put together the most fertile human minds and ask them to set to work generating a list of possible reproductive behaviors and techniques and their list would represent only the most meager selection of gray tones in comparison to the colorful reproductive palette that has already been discovered by biologists who specialize in the biology of reproduction. The world of reproduction offers a painter's palette of every color, hue, chroma and value imaginable. The art of reproduction is more complex, deliberate, delicate and beautiful than the finest that DaVinci, Titan, Tintoretto, Picasso or Guayasamin ever produced.

Adaptations to temperature and environmental control are equally varied: dark hairs to trap heat, light colored hairs to reflect light and reduce temperature; hairs to reduce wind evaporation, no hairs. Some sap and juices don't evaporate or freeze; there may be high or low surface area: volume ratios that reduce, increase or prevent heat loss; leaves of every shape and color each have advantages and disadvantages. Individuals vary in the genetic expression of their DNA code depending on the environment they find themselves struggling in. Change an organism from one environment to another and as far as it can possibly do so, it will change its properties so as to take advantage of resources in the new environment as best as possible; phenotypic variations based on the genotype, the environment acting on the gene expression and the genotype influenced to mask, hide or reveal possibilities according to the environment. What a dance.

We could continue on and on elaborating upon the range of variation in organisms but you would soon become bored, if you are not already, and I would get to do nothing but tap out words at the computer for the rest of my life. Neither is desirable.

Let's momentarily go back to our theoretical island and its array of rich and diverse environments. We'll leave this chapter not by painting any one picture of the island but rather by asking ourselves a few rhetorical questions. Depending on where our island is, how could organisms reach it? Who and what organisms would arrive, colonize it and find it acceptable, and who would land there and find it not to their liking and depart or die? When and what would they do when they got there and how would they capture the necessary resources to survive? What adaptations would be advantageous in each of the differentiated environments and which would be non-advantageous? The questions are endless!

In science we frame these research questions as hypotheses and design experiments to test and examine each one. Careful experimental design requires that each question have a null hypothesis (Ho) and an alternate hypothesis (H1) which we test with carefully and truthfully collected and analyzed data. If we reject the null hypothesis then we must accept the alternate hypothesis. More on the scientific method later but …the scientific method instead of reducing the number of questions we have to ask of nature actually increases the number. The more we learn, the more we learn about how little we know. In science, ignorance is a kind of bliss that generates wonder and curiosity and teaches us humility. On the way to answering the questions some useful and practical things are elucidated that increase our adaptability and let us manipulate our environment to suit our ends.

The biotic potential of our theoretical island is truly immense but what ultimately will find a home on it and for how long it will make a home there is a topic for a future foray. In closing let's take comfort, or find some curiosity, in the fact that even Bikini Atoll has been re-colonized. The effects of the atomic bombs that were detonated there during the past century couldn't prevent life from working around the problems of heat and radiation. Life prevails everywhere even under the most adverse conditions – it may be temporarily eliminated but seldom for long.

On Species Change in Ecosystems

When several or more individuals of different types of organisms find themselves together in a particular area in pursuit of their life needs we refer to them as biological communities; several species all trying to get their needs met on more or less the same piece of real estate at the same time in history. The communities together with their environment and the energy entering the system from the sun are referred to as ecosystems. Ecosystems are often spoken about as though they were real, concrete things in nature akin to an individual. They are not and the application of the term is really quite mushy in our communications with each other. Some people refer to the earth as an ecosystem, others to grasslands, temperate or tropical forests as ecosystems and yet others to the city as an ecosystem or a drop of water full of life. All of these are acceptable usages of the term, providing the user defines the term and the criteria of usage.

Language is imperfect in describing even simple things and it is most exact when there is a common agreement among all parties about what a particular word means. When there is not a common understanding then miscommunication arises and imprecision results. Take for example the word chair and the concept it describes. Most of us generally have no problem with the word because there is a general understanding amongst us about what a chair is. To classificationists, linguists and philosophers however, 'chair' does pose problems because not all chairs have legs and a back; some are soft globes filled with Styrofoam chips, others have four, five or six legs and not all have backs but we still refer to them as chairs. Pretty much all chairs are things you sit on, however chairs are not the only things one can sit on. One is left with the problem that if they sit, for example, on a table, does the table become a chair? 'Chair' describes a concept, an extraction of common properties from the collective population of objects. The more precise and predictable the common attributes are, the more precise the word and consequently the less opportunity for misunderstanding during communication. Insofar as possible one wants to define what are known as the necessary and sufficient conditions for the usage of a word: *only* what are necessary and sufficient; no more, no less. It is tricky and serious business this classification stuff and much more important than most people think. Generally our language, imperfect though it may be, is good enough for most communication unless of course you are trying to prepare a legal document in which case you had better give the words some serious thought. If your mind has a particular bent in that direction, trying to define the meaning of a word precisely can provide a bit of fun and diversion during idle moments.

People who classify and describe organisms or groups of organisms have formal International Rules of Nomenclature that they must follow but there is no common rule book for classifying ecosystems. The term ecosystem is generally understood to refer to a situation where solar energy, abiotic substances, autotrophs, heterotrophs and decomposers all are present and work together in some form of interrelated system. There are many variations and elaborations of this concept and how and when it should be used but it is not my intention to struggle through them. If you take the time to look up the meanings of abiotic, autotroph and heterotroph and all the various attributes referred to by these terms you will get an idea of how complex the problem of defining ecosystems is and how much room there is for misunderstanding. Fortunately our language is structured so that adjectives describe and elaborate on nouns and adverbs help to define an action. The easiest way to reduce the ecosystem miscommunications is to elaborate on our nouns and verbs and refer to forest, grassland, intertidal, aquatic and so on as ecosystems or if we want to be more precise we can refer to pine forest ecosystems, fescue grasslands or more precise yet; Rough Fescue or *Festuca scabrella* ecosystems.

The level of precision with which we define nature depends on our purpose and there is no limit to the degree of precision we can shoot for if we have some motive or reason to do so. A good deal of the human scientific endeavor is dedicated to being precise so that communication is clear and ideas and experimental findings can be tested and repeated by others with similar interests. If you conduct an experiment and can't repeat it again to see if the outcome is similar, your findings are at best inconclusive and at worst worthless. Peer testing and review are important in science as general principles are sought.

Recognizing that 'ecosystem' is a general term, let me begin by saying that at the level of general synecology, natural history and resource management, I will refer to ecosystems as recognizable grouping of species that are generally found together on one or more recognizable types of environments and that also generally repeat themselves over a landscape on similar environments. Furthermore, in naming and placing boundaries around them there is some merit to using either the dominant environment(s) they occur on or the dominant and easily visible plant species that characterize them. Naming them after the dominant animals doesn't work all that satisfactorily because animals have a habit of walking, flying or swimming around and it is hard to place boundaries around such activities and the areas they occur in. Also there tend to be many fewer animals than plants and the big recognizable ones are not always visible when you want to place a chunk of real estate in one category or the other. Plants work well for simplifying, defining patterns and placing geographic boundaries around bits of nature. They are not immensely precise but they generally work; they are 'good enough' simplifications of a complex world and sometimes one has to settle for good enough until something better comes along. A 'grizzly bear' ecosystem I don't understand but a 'white spruce' ecosystem with a lot of *Equisetum arvense* that grizzly bears use I can more or less get my head around.

This game of ecosystem classification isn't all that easy because ecosystems, like all things, change in space and over time. Generally in highly diverse and variable terrain the physical and chemical properties of the landscape change quite quickly and over relatively short distances. In these situations the plants that grow there generally form easily recognizable patterns and groupings as they finds bits of the variable habitat more or less agreeable due to their particular needs and adaptations. Boundaries are recognizable and understandable. In environments that are more or less flat with relatively uniform soil and climatic conditions the ecosystem boundaries are much harder to recognize and in these areas the change in plants across the landscape is much more gradual. One group of plants gradually grades into another over large distances and boundaries are nearly impossible to recognize.

In both highly variable and highly uniform environments when one places a boundary they are making a decision concerning recognizability and boundaries in nature are never absolutely precise. There is always an area with a mixture of species common to both communities; in ecosystem classification we call such areas ecotones to imply a mixture of system properties between one or more generally recognizable types. Such is the nature of ecosystem change over space. The concepts and their application are easily understood but their application takes some practice.

Ecosystem change over time is another problem all together; much more fraught with intellectual minefields, pitfalls and traps. Tough business, this studying of ecosystem change over time. That nature and her ecosystems change a lot over long periods, we are all aware. Over time on a geological scale the earth has had a vast array of environmental and species changes. The Devonian period had immense diversity of life in the seas of the world and a flowering and flourishing of the bony fishes; the Triassic, Jurassic and Cretaceous periods enjoyed an immense diversity of dinosaurs and ferns. Since the end of the Cretaceous period nature has made some wonderful explorations into the endless possibilities for flowering plants and mammals. Along the journey many species got wiped out, eliminated, made extinct because the environment changed and they found their particular set of characteristics badly matched to the new conditions. Extinction and evolution of species is a natural process but thanks to coding errors new forms of life arise to replace those that no longer work.

Similarly environments change immensely over geological time; the central plains of North America used to be inland seas that during some periods stretched from the Gulf of Mexico to the Bering Sea in the arctic. The Rocky Mountains used to be under water and part of the sea bed until the huge pacific plate began to collide with the smaller North American Plate and sub-duct beneath it. Heavy things push and deform light things and the small North American plate began to buckle and contort and the Rockies started to get pushed up; they continue rising to this day just as do the Himalayas and the European Alps; they are all young mountain chains. Mountains, when they are no longer getting shoved up faster than the forces of weathering and gravity can bring them down, erode and get smaller. The Earth is littered with shrunken remnants of former tall mighty mountains; the Laurentians, the Adirondacks, the Atlas'. Likewise the Earth's climates change radically over time. Periods of climatic instability have been the rule and not the exception; periods of warming, periods of cooling that resulted in the advances and retreats of glaciers over much of the northern parts of our globe and changes in sea level as the water from the water cycle was incorporated into the ice on land and sea. This glacial period only started a blink of an eye ago in geological time and it only ended about ten or twelve thousand years ago. In fact it may not even have ended and we might still be in the glacial period; nothing more than the Earth taking a pause in its continual struggle to make peace with global energy balances.

Humans don't have a particularly good knowledge of the global dance of large cycles in nature because the cycles are quite complex and take place over long periods of time. We, because our life perspective is only sixty or eighty years if we are lucky, tend to be clouded and biased in our view by our life span. Perhaps we would view nature in quite a different light if our life span were a thousand years instead of sixty to eighty. I suspect that when you live for a short time you see things differently than if you were privileged (or doomed?) to live for a long time. Perspective is important.

Over geological time physical and chemical environments change a lot and species and ecosystems likewise change a lot. Change in nature, some of it extremely radical stuff, is the rule and not the exception and all species must adjust to the change or walk away from the game table. No species has ever been, and probably never will be, able to stop the big cycles of nature; they are just too mighty forces for puny organisms to play much of a role.

Organisms, however, are good at controlling, adapting to and changing the smaller cycles of nature and manipulating local environmental conditions; birds build nests, gophers slip underground during cold spells and mud puppies bury themselves deep into the slime during periods of drought. Some simply walk away from their environmental problem; some birds go south to warmer climates in the winter, as do caribou. Each species tries to optimize its advantages by adapting to or changing conditions. The result is ecosystem change at a much smaller and more local level. Let's explore this change.

We'll begin just as we did when we talked about the formation of the volcanic island in an ocean and how the environments became differentiated by physical and chemical processes. We'll establish some assumptions, build a model and see where it takes us. After that we'll briefly test it and see if it conforms to reality in a reasonable way. Models are not much good if they don't work well enough to be at least useful in some way or other. The model must adjust to fit reality because reality certainly won't adjust to fit our model.

Let's imagine that we get hold of a grader and with the help of this powerful technology we scrape away all the plants, animals, microorganisms, and organic matter. We'll strip it right down to the bare geological material and we'll make it relatively flat because processes on flat and more uniform areas are generally simpler and more easily understood. We'll be the proximate causal agent of environmental change instead of waiting for an earthquake, volcano, landslide, glaciers, fire or flood to do a similar job. After we have done this we will get out our chair, sit down, watch for an awfully long time and see what happens. In fact, while we're at it why don't we keep several, or several hundred, notebooks at hand so we can write down what we observe. The first note we should make is to write down the date and time when we finished our grader work; let's call it Time Zero to note the moment when the resources in our area were available for life to get a shot at capturing.

Before we have even had a chance to sit down, the biotic potential of our area has set to work and all manner of living things have begun to invade it. Microorganisms blown in by the wind, spores and seeds of many species have arrived; insects have flown over it and perhaps defecated and deposited organic matter along, possibly, with some seeds, spores or other types of organisms that have passed through their gut. Other insects and spiders walk into the area and leave their outputs like the flying insects. There is not a lot of food at this stage of development so our area gets predominantly outputs that organisms accumulated from other more elaborate neighboring biological systems. Rains come and wet the area washing salts and other chemicals from the upper reaches of the land surface down the soil profile and deposit them at depth; the sun dries the soil and the winds bring in dust and organic matter from outside the area and the area loses dust from within the system. Our system experiences material gains and losses of all manner of physical, chemical and biological products by all manner of fundamental natural processes.

Eventually some of the plant seeds germinate and push their roots into the soil and their leaves into the air for seeds germinate even if the site does not have the light and nutrients they need to survive and become independent; the seed has enough stored to get it started in the new environment for that is the way seeds are – "just add water". The young plants of some species, if they have suitable adaptations, find the site at least minimally acceptable and grow and become independent. These self-feeding autotrophs are able to capture enough light, enough carbon dioxide, water and essential nutrients to get by. Others start out only to find that they had fallen on unsuitable ground and they die leaving what ever organic matter they had on the land. Those species that found the site acceptable continue to grow and push their roots deeper into the soil, loosening it and providing nutrients for some of the root fungi and other microorganisms that came to the site. The above ground parts of the plants provide shade over the soil and reduce water loss from evaporation in specific areas. Leaves die and are decomposed or are eaten by herbivores of all sizes and shapes which in turn deposit dung and urine on the site and provide nitrogen and other chemicals to the soil. Some of our herbivores are eaten by carnivores and some carnivores are eaten by other carnivores; all in their turn die or are killed and bits are left to decompose and break down. Some of the microorganisms increase in their abundance quickly, perhaps duplicating every ten minutes or so; others take longer but the biomass of our area increases bit by bit, slowly at first but more rapidly as the system develops. Plants push their roots deeper and wider in the soil as they grow and need more water and nutrients to support them. They produce flowers if that is their way and birds and insects come and pollenize them and more seeds of more species are deposited in our site. Some of the species that have come to our site have come from nearby, perhaps most, but some will have travelled vast distances with the help of the wind, water or other animals.

All manner and abundance of organisms have migrated into our denuded area; some have found it acceptable and made a home there and others failed. All, in their attempts, have contributed to the development of the site and all, as a result of their efforts, successful or not, have changed the site. The soil has organic matter now, it holds moisture better, has a different nutrient composition than it did initially and some areas are shaded more than others and yet others are more exposed. Plants and animals that like more or less shade have found a place in our rapidly changing ecosystem. The fundamental ecological processes worked on our site as they do on all sites all the time; they can't be stopped any more than commanding the waves to cease will calm the water.

Let's stand back a minute and look at our site from a slightly more distant perspective. Belly biology gives a different view of things than armchair biology; hilltop biology provides yet another perspective. We will note that there is a pattern beginning to form on our site. Those plants that spread by runners, rhizomes, tillers, or suckers occur in patches because they have had a particular advantage in the space game; they have been able to tap the nutrients from the well-established mother plant who has her roots deep enough in the soil to supply the goods and services, water and nutrients, necessary for horizontal and vertical growth. Those plants that produced seeds have dropped them nearby or dispersed them a bit farther afield and new plants of each species have started and begun to colonize and capture new areas of the site. Some seeds have left the system and form part of the biotic potential of areas beyond. Our site is a wonderful array of color, pattern, tone, shape and size. There is a lot of biological diversity on it and it is not distributed evenly; plant and animal species are grouped but recognizable patterns have emerged.

In the process of filling up the site with life, the physical and chemical aspects of the site have also changed. The soil has become more acidic from organic decomposition and the more acid soil has changed the mobility of other chemical species and these get moved to greater or lesser depths depending on their mobility. Some heavy metals get moved to greater depths. The soil changes as a result of various additions, removals, transfers or transformations. The roots of plants have also drawn nutrients from deeper in the soil and when the above ground plant parts die these are deposited on the surface; we call this biocycling. These surface nutrients are then available for plants with shallow roots to capture and use. The organic matter and the shade have differentiated the environments on the site even more and also changed the climatic conditions. Areas with shade and organic matter are cooler and don't dry out as quickly. We will find that our pattern of species distribution reflects these changes.

Over time the changes to the environment as a result of our organisms are so great that the changed environment is no longer suitable for those organisms that originally colonized it; they can't get what they need. They are initially reduced in their abundance and gradually die and are eliminated from our site. New species better adapted to the new conditions gradually migrate in, colonize and make their home there as long as the conditions are suitable. They in turn take from, give to, alter and transform the site to initially capture as many resources as they can and prevent other organisms from getting them. Once light, space, water and nutrients become in short supply the competition gets fierce in the war for meeting needs. Those plants and animals with the best natural technology (if I can use that word), the best mechanisms for getting resources, get the most resources. The best adaptations are the ones that win in the resource battle but our site has so much differentiation that no one species can capture all the resources; lots of species diversity from the large to the small exists on our site. Those species that were generalists find that they were good initial colonizers but as the site changed other species with more specific skills were better able to take the resources from them. The process of site differentiation continues and the resource battle wages on; competition gets intense and the system reacts to the competition; some win and some lose and the system continues to change. Highly differentiated sites favor higher degrees of specialization and higher levels of specialization favor higher degrees of diversity in an interesting loop. This process will continue until our site has undergone enough change so those species that are able to hang on get their piece of the pie under the prevailing conditions, find a balance and settle down to the process of growth, death, reproduction and replacement of each generation.

Gradually, ever so gradually, the site reaches some kind of balance as nature tries again to reconcile resources and energy. Inputs and outputs become a bit more predictable and the system, like my garden pond, does not experience quite so wild swings as it did initially. Stability, of course, is never reached because temperature, light, moisture, and nutrients vary in their intensity, concentration, availability and mobility as the seasons come and go and the big cycles of nature wreak their havoc. Some kind of dynamic equilibrium seems to become established and the diversity of species is high although the abundance of each species is reduced.

Nudation, migration, competition, reaction and stabilization; five simple words to summarize the highly complex processes we have documented. A nice orderly and tidy piece of theoretical description that at some level seems to work; the model fits reality. We feel proud of our observations and get a sense that we understand nature. We know what will happen. Initially the system experiences a directional change from start to finish and gradually the directional change will give way to non-directional replacement change as the system stabilizes. Progression gives way to stabilization. Biomass increases, biodiversity increases, specialization increases, complexity increases and the resources get cycled through many different creatures and used efficiently the longer the system develops according to its ways. We get ready to put our notebooks away; the dependent factors have come into a dynamic balance with the controlling factors and the ecosystem house is staying together. The experiment is over.

Before we fold up our chair, however, a fire, flood, tornado, landslide, pest outbreak besets our test system or perhaps, our neighbor comes over with a grader ... the possibilities are endless and the upshot is that our stable little system gets disrupted to a greater or lesser degree and changes. The toothpaste gets squeezed in the middle of the tube and things get upset. We unfold our chair and sit down and watch again.

Depending on how much disruption the system experiences the system changes back to something that is a bit more similar to one of the early stages. Species that we had observed early on but were eliminated start to come back in, thanks to the vast biotic potential of our area, and re-establish themselves. They may not occupy exactly the same spots as their predecessors but many return; plants, animals, microorganisms. The system seems to have regressed to an earlier stage of development and any reasonable person would recognize this as nothing more than a temporary setback; things are predictable and we won't waste any more time watching because we know what will happen. Just then we notice a species or two on our disturbed site that we had not seen the first time around. Chance played a role and the biotic potential of the area threw our site a new curve; some species with different needs, different adaptations, different skills. Chaos prevails in our system for a while and we have to rethink our model and allow for chance. More watching and we note that the ruffled feathers of our system get sorted out, re-groomed and the system once again starts to progress to follow the same general processes as our earlier one only this time with a slightly different complement of species. In time, we note, the system develops to resemble what was there at the endpoint of our first set of observations. Resembles it only but this time with a few species added or subtracted, most species in slightly different positions with slightly different abundances. Similar but not the same; close enough in our classification to be considered and described as the same ecosystem. Thank goodness it is close enough to what we had before because 'we really would hate to see nature destroyed' and something different develop. We breathe a sigh of relief. The world unfolded as we wanted it to and in accord with our sense of order. Our theory holds and our predictions came approximately true but not quite. The general prediction was accurate this particular time but not overly precise; we didn't get the prediction exact; we missed something. Perhaps we didn't understand chance and chaos well enough. However, for most operations the prediction was close enough.

Climate, geology, topography, hydrology and biotic potential all act in their fullness over time on a particular bit of real estate to produce a particular ecosystem that will last there for an unpredictable length of time. All acting on all bits of real estate on every square centimeter of the globe to produce all the different environments, species, variation, amino acid base pairs, communities and ecosystems in an ever changing collage.

On Simple Mathematics and Ecological Understanding

Somewhere along the continuum we humans developed mathematics and it has been a useful tool in expressing concepts; no less in ecology than in other fields. One of the useful mathematical concepts is the notion of the equation as it implies a balance on both sides of the equal sign. Let's briefly try and put our description of change into a mathematical expression, not so we can solve it and reduce it to numbers but rather to see if we can express what we have learned in a simple way. Who knows? It might prove useful to we humans as we go about our business of trying to survive and capture our needed or wanted resources. Perhaps we humans discovered the principle long ago.

Let's start by acknowledging that when we first graded our piece of land we established a particular climatic, topographic, hydrologic and geologic regime on the site and since we didn't deliberately bring in any particular species of plant, animal or such to the site, the site simply experienced whatever the biotic potential happened to be at that moment in time. Since this is what it was immediately when we parked the grader we will call it time-zero for it was the initial state of the system when system development began. These attributes to a greater or lesser degree will control what follows so we will call them controlling variables and we'll stick them on one side of the equal sign of our equation. All the changes that occur in the system and all the properties of the system as it develops are, to a greater or lesser degree, dependent on what the initial state of the system was like. So we'll plunk these on the other side of the equation and say that at any moment in time the vegetation, microorganism, animal, climatic, geological, hydrological, topographic (and on an on) properties of our system occur as a result of the interactions among the controlling factors. In fact instead of listing all of the possible things that go into comprising the properties of the system at any point in time we will simply call them present ecosystem properties and be done with it. Since it is only a shorthand expression we can easily expand, contract or alter the list of present ecosystem properties as it suits us for our particular enquiry. We can throw in social, chemical, psychological properties or anything else we want. After all everything is included in the definition of the ecosystem. If we do this we will end up with an equation that looks something like this:

Present ecosystem properties = function (climate, geology, hydrology, topography, biotic potential, time)

Now equations are interesting in that they must balance. The left side must always equal the right side. If the sum of the left side changes then there must be a corresponding change on the right side and visa versa. If only one of the variables is changed on either side then the other side must also change to keep in balance. Interesting thing about equations though, you can change either the left side or the right side without changing both. One side can be reshuffled as long as the ultimate resolution of that side does not change. For example we can say that 3+4+7=14 or we can say that 7X2=14 (or any one of numerous options). The left side reshuffled and the individual numbers changed but they still worked out to 14 so the right hand side didn't have to change. Not every change on one side or the other necessarily results in a change on the other.

Interestingly if one knows what changes take place on one side or the other they can predict what range of possible changes might occur on the other. At least they can give a range of possible outcomes. A level of prediction is possible but one can never say with immense precision exactly what the final configuration of the equation is until the outcome is resolved. Particularly if one side has a random number thrown in for good luck. Chance has been introduced into our equation although in all fairness truly random numbers are hard to come by in this world of mathematics. If the computer generates them it usually does so with the assistance of some particular algorithm that is predictable and hence not random; if a person calls out the first number that comes into their head, that person has a predisposition toward favoring one number set or another. True random is scarce if not non-existent; there is some level of order to what we think of as random. It may appear chaotic but even chaos seems to have some level of order.

Ecosystems seem to work a little bit like our equation with every change being balanced by a corresponding change either on only one side or both sides; depending if the sum of a side changes or not. Even the 'random', or perhaps semi-random events, like our introduced new species, our landslides, floods, or fires occur in our ecological equation. We do know enough about risks of pestilence, fire and flood to know when there is an increased likelihood of the event occurring. Ask any insurance company statistician who generates the probability numbers upon which our premiums are based. Life experience is a great teacher in the game of risk assessment. The young are generally more experimental than the old partly because they are better able to cope with failure but also partly because they are not long enough in the tooth to know the world is a dangerous place. High risks, while they may generate great rewards, increase the odds that one will be eliminated from the game.

Sometimes the amino acid base pairs get reshuffled in such a way that a great new path across the tides of history is revealed; more often than not, experimenting individuals go bust and can no longer play at the table. This is rarely a problem in nature because there are always new players coming to the table. That is what reproduction is all about. Resources are linear mathematical functions while the outputs from reproduction, when all the individuals of a species get in a frenzy about it, can become close to exponential. Unlimited individuals chasing limited resources. The equation must balance.

On Us

Somewhere along the time line of recent history, perhaps a million years ago or so, a unique species arose that was a tool user of grander proportions than the other tool using species. We humans, US, didn't just use spines off plants to poke food out from shells, spin nets and webs to capture prey or other clever and useful acts in the struggle for bringing resources from the external environment to the mouth so that what we don't use in the gut comes out the other end. Our species began to use tools in a much more elaborate and complex way. We didn't simply take bits of nature and use them as they were, didn't simply dig holes or build nests to live in as a way of modifying our environment so that we could live in places that normally would not be habitable. Many species do such things and I'm not slighting the wonderful and elegant nest and burrow diggers and builders because they and their creations truly are marvelous. They do not seem to record and pass on knowledge as do we humans.

Furthermore, like many other species, we are a social animal. We like to hang about in groups and we like to share our time in the company of others of the same species. Being a social animal does not make us unique for there are literally thousands upon thousands of social animals, many with highly elaborate social structures: bees, ants, lions, jackdaws, chimpanzees. Elaborate social structures are to be found everywhere in nature and in most of these societies there is a social hierarchy with different individuals having different places and roles in the various societies. Quite wonderful! Complex individual organisms work together in generally cooperative and generally advantageous ways. Feathers and fur get ruffled between individuals or groups in most societies but in most organisms when the things they build get destroyed they seem to be reconstructed in roughly the same way they were before. Knock a robin's nest down and it will be reconstructed in approximately the same way; sweep away a spider's web and it is replaced with a structure that is much the same as the one the broom got hold of.

We seem to be different in a number of ways which taken together make for an interesting species. Most of our individual attributes are represented in various species; many species build homes; many use tools, form complex social structures, cooperate, show aberrant and unacceptable behavior, develop communication skills, or even seem to exhibit the complex emotions of sadness, joy, pain or laughter or any of those things characteristic of humans. Probably no other species however, has been able to wrap up all these individual characteristics in one interesting bundle and then throw in a few other intriguing characteristics just for good measure.

What are some of these good measure characteristics? We are extremely and intensely curious and experimental. Different individuals and groups of our species look at the world in different ways. We have empathy; develop complex games that are designed simply for pleasure, with gaining life skills as a secondary goal; seek entertainment and entertain ourselves by watching how other species do things and then if it so pleases us, we imitate other organisms. We perhaps, are the ultimate mimicker; we shamelessly copy and steal ideas from anywhere we can observe them.

Apparently no other species has taken more or less thirty complex, nonsense sounds and put them together to describe and communicate what we observe. No other species, as far as our observation tells us, has a communication repertoire that includes hundreds of thousands or even millions of words to describe equal numbers of abstract concepts. And that only includes verbal and written communication; add into the repertoire, communication by smell, gesture, visual, vocal and aural means and the communication options are much grander. Painting, sculpture, music, writing and dance are uniquely human because we undertake these displays to provide pleasure or express what we observe and don't simply do them to help the mating or food-getting processes or to capture and control space. No other species has taken these complex nonsense sounds and represented each with some kind of glyph, a picture, and then put it all together into a complex written or drawn form of communication. Others have not developed a system of counting, adding, subtracting, multiplying or dividing and then put these together to form equations to describe and explain our observations and no other species seems to have developed a complex system of measurement to describe time and space and then used it shamelessly to modify its environment.

Those species that are objects of our study are, for the most part, passive subjects as they do not understand the senior game of human intellectual enquiry. Some in true Pavlovian form, may well offer us the action we want, or expect, in order to get the reward we offer, but sheer intellectual inquiry for the pleasure (and perhaps profit) of it, does seem to be a uniquely human endeavor.

We have a sense of the past, the present and the future; a sense of order, disorder, right, wrong. A sense of place in the order of things and a feeling that we are part of a grand experiment. No other species, as far as we know, has a concept of a God, yet worship and religion in some form or other seem to be ubiquitous throughout all human societies dating back to early times. We have a sense of goodness, truth, beauty, right, wrong and morality and their opposites and we explore these ideas and our world with vigorous enthusiasm and curiosity. We have a hot fire in our bellies.

Not only do we exhibit all of these characteristics, we know we have these characteristics and we celebrate or lament them to a greater or lesser degree. We are sentient, we have knowledge, we are conscious of this fact and we are self-conscious. Furthermore we have a horizontally opposed thumb found only in a few other species and this is one of our greatest physical adaptations to the environment. We use it to great advantage.

We are also an unpredictable, plastic, very adaptable, highly variable species and we colonize and live in most of the earth's diverse environments – all year long. We change and adapt probably more quickly than any other species and we have developed, as a result of all our special characteristics, the ability to respond rapidly in highly variable and innovative ways to different situations. Our concepts of what a resource is change as we learn and we have turned natural things into useful resources that no other creature has even explored; coal, oil, uranium, money, fire and gold. Resources are discovered in the minds of men and women!

We are intensely arrogant, intensely humble, sometimes kind and generous, and other times quite brutal. We are always self-involved. Probably no other species can take small errors in their genetic code and find opportunity where disaster lurks and certainly no other species deliberately manipulates this genetic code in other species as we have done for thousands of years of experimentation in animal and plant husbandry. We are intensely competitive with each other yet we deliberately share some of what we have with other individuals of our species that we have never even seen or met. Compassion is an important human attribute. We generally have a sense of the public good even if our individual acts are contrary to it. Guilt hangs around in all but the most sociopathic soul and perhaps it is even lurking there. We often undertake acts of altruism and frequently act against our own best interests. Often we are too stuck in our own ways, too rigid in our approach and oftentimes we can't see the forest for the trees. All of us!

From the perspective of ego and self-importance, there is no species so richly endowed as we humans. We truly believe we are both the cause of and the solution to environmental degradation, whatever we choose to define it as. We, at least some of us, are so self-important as to presume we are above the laws of nature and that we really are of key importance in the balance of natural things. One does not have to look far in the history of human endeavor to discover the fate of high art, high culture and high pursuit. Civilizations that caused significant and high levels of environmental changes have been brought low, buried, sedimented and cloaked in a mantle of native ecosystems only to be discovered centuries or millennia later as a result of human curiosity.

In short, we are without a doubt the most interesting species and, given our arrogance and self-involvement, we generally are most curious about ourselves, our past and our future, than we are about all other species and attributes of this world. Insofar as we can, we manage the world to suite ourselves and our current notions of what we want, what we value, what is right, wrong or necessary. These management goals and techniques vary over time in accord with our tools and our values and we worry about some things more than we should and other things less than we should. We are innovative and co-operative specialized generalists.

For the rest of this story let's examine ourselves and how we relate to our environment. We won't do it in a very comprehensive way for the story is too long, too complex and too convoluted. Besides, I get bored easily with the path I have travelled and feel the need to get on with something different. There is much I have not looked at, I have left many stones unturned and the time is coming when I must step away from the gaming table and become a resource to be recycled and used by those creatures that follow. That is the rule; you can only play for so long and one never knows when and for whom the bell tolls. It might be you; your future is always uncertain and increasingly so as time passes.

On Hunters and Gatherers

Deep in our souls at our very core, we are, I'm convinced, all still hunters and gatherers. In developed nations, those individuals who chose to participate in the fruits of the industrial economy are predominantly metaphorical or symbolic in their pursuit of hunting and gathering but some still enjoy the chase. They don't need to hunt and gather but they choose to.

In countries that still allow guns, and bows and arrows, it is a ritual for those that so choose, to go out in the country in the fall and shoot a few ducks, upland game or perhaps a deer, moose or some other creature. It gets you to your roots and pits you against other species that constitute prey and food. You must be crafty, silent and stealthy for these animals are most certainly not like cows standing in a field. To be a successful hunter you must study nature, you must learn the ways of the animals and listen to the other species that give hints of movement; the sound of a squirrel, a bird or perhaps the snapping of a twig or the bellow of the prey itself. You have to keep your eyes peeled and your ears well tuned and alert; even your nose can be helpful.

Years ago when I was a young fellow, and at that time a hunter of birds, it used to amaze me how loud the beat of my heart sounded as I stood at the edge of a small forest clearing where the berry bearing plants typical of disturbed clearings in forest ecosystems flourished. There wasn't much Ruffed grouse food to be found in the dense forest so I learned to stand at the edge of the clearings and watch and wait for the movement or the sound of movement from a snapped twig or the rustle of the leaves as a body rubbed against them. It was a tense moment and required absolute silence on my part. The grouse had gone quiet because of the noise I made as I approached the clearing. I had learned that if I remained silent and motionless, ever on the ready, that eventually the grouse would conclude that the danger had passed and begin to go about its business; I would have my chance and if successful we would have a tasty grouse meal enhanced by some wild mushrooms, a bit of bacon from the local farm and perhaps some pie baked from locally gathered saskatoons or huckleberries. I learned that certain types of ecosystems were more attractive for Ruffed grouse and that their use differed throughout the day; the clearings that had lots of the choice berries at feeding time; the edges of roads where there was exposed gravel for the crop were important at times; and then there was the forest where the birds would roost in the evening or early morning.

Grouse hunting was a good teacher when I was young and I always felt a certain pride in showing and sharing the fruits of my labors. Many of the areas where I used to hunt are still rich in Ruffed grouse; some have changed, unhelped by human hands, because ecosystems change over time and the early successional vegetation that had been established due to some disturbance factor has become dense forest and no longer suitable for grouse. Other forest areas have been disturbed since then by fires, wind throw, clearing and the berries are abundant in these areas now. Some of the areas I used to hunt have fallen prey to the developer's hand as they have been turned into housing estates to accommodate the ever-increasing human population. There were only about two and one half billion people on earth then and now we are approaching six billion. I don't hunt any more but perhaps ... just for old times sake. The memories are sweet and there was a certain sense of being fully alive during those activities.

I suspect my experiences were not unique for I have heard many hunters tell stories since then and what they describe is similar. Most have a strong reverence and respect for their quarry and most love to get out in nature and away from the pressures of the daily grind in the industrial work world; sometimes the sensitivity is masked by a bit of bravado. There is a certain refreshment and rejuvenation to the whole ritual. Most hunters I know are ardent conservationists and willingly donate time and money to ensuring that the land remains productive for their chosen quarry. Most are not violent people and most are cautious as they are aware of the power and risks associated with their weaponry and the vast majority, all but a tiny number of them, are scrupulous in abiding by the hunting regulations for they appreciate their importance in ensuring that over-hunting does not occur.

In those countries where hunting and fishing are non-essential and pleasant past times there are usually a lot of laws governing them. No hunting during breeding or birthing times, only so many animals may be taken, hunting only during certain hours and a host of other controls over equipment and activities. Humans have studied the populations and we have a general idea of how to maintain our desired population balance. Sometimes we want more of the quarry than unmanaged nature would produce and we alter the ecosystems a bit to maintain the desired population balance; giving it a nudge here and there to change the ecosystems so that they are more suitable for one species or another. Rarely do species that humans value become extinct for they are monitored, carefully tended and controlled. Fishing is similar to hunting in both gesture and experience. They are important economic activities and the fruits of that part of the economic cycle related to them get recycled and redistributed to other areas of the economy; parts go to supporting social programs, hospitals, and other bits of the infrastructure of developed countries.

Wealth is important because it generates surplus and reduces the need to take care for the 'morrow. If the hunters and the fishers come home empty handed they can always slip into the local grocery store and pick up something for the plate that has been produced in a much more intensive and productive way. Something that is the product of intensive ecosystem management where undesirable species have been eliminated or controlled and the bounties of the sun, air, water and nutrients directed toward the maximal or optimal production of the favored species. A cost benefit analysis would show that hunting and fishing don't make much economic sense; it is cheaper to buy in the store, but wild food tastes different. I would be remiss if I failed to observe there are those who claim a cost benefit analysis would show that meat eating does not make sense. Be that as it may, most humans given the chance, eat animal protein!

In those places where hunting and gathering are not pastimes and are pursued out of need, and the drive to meeting basic needs, the story is quite different. If you rely on nature to provide on a daily level then you must withdraw from nature on a daily level and hunting and gathering seasons are less of an option. The failure of the hunt or the root and fruit gathering exercises causes genuine hardship and want unless you have put away surplus for the future tough times. Storage of surplus food is a tricky business because rarely do the decomposers in ecosystems miss out on the opportunity for the goods they need. Food storage is a big problem and a serious business for obligate hunting and gathering societies. Many have developed elaborate technology, relatively speaking, to preserve; drying, salting, cooking, smoking, placing in containers and sealing them, burying or freezing them.

There is a small Inuit community in the Canadian north that had a community pingo, one of those ice features that rise above the general landscape for ten or fifteen meters, have an ice core and only last for several years before collapsing as the ground gets stretched too thin as the land rises and the sun melts their exposed ice core. The community hollowed the ice core out and used it to store their food during the summer months. It was an effective refrigeration technique and as far as I know it is still in use.

When doing some research a few hundred kilometers west of Tuktoyaktuk our party had the help of an Inuit research assistant. It was during the hot arctic summer and our assistant took his gun, our freighter canoe and set off to get a barren ground caribou. He returned shortly with the hindquarters and explained that he had left the rest for the bears. I was curious about how the meat would be kept fresh in such a hot area with an abundance of insects just waiting for a chance to get at our meat. Electricity and refrigeration are a part of my world so my ignorance of arctic survival was obvious to me. I watched and listened. Our fellow had not skinned the hindquarters and he got a rope and tied them to the eaves of our cabin so they hung in the hot sun. I soon learned that by not skinning the quarters, the hide prevented the blowflies from accessing the meat. The exposed cut, hanging downward collected blood which quickly, in the hot sun, dried and sealed off the meat with a hard crust and prevented the flies from penetrating and spoiling the meat there. We fed off those hindquarters for the best part of two weeks, slicing the needed meat off the exposed cut and letting it reseal. The quality was excellent and it never went rancid. The culture, over time had found a way to preserve nature's bounty and the skill had been passed down through the oral tradition and by country education over the years. The old passed on the skills they learned in their youth from their parents, to their own young.

I said earlier that I enjoy sailing. Each year, when able, several of we fellows slip off to the Canadian west coast for a week or so of sailing just to give our wives a break from us. We are generous and understanding husbands, as you will appreciate. One of our sailing colleagues had spent many years in Newfoundland and had learned the ways of food preservation in the remote Outports. He decided that we should enjoy a breakfast comprised of standard fare in these areas and informed us we would be having Brews; strips of salted and sun dried cod that had been soaked to make it soft, all placed over top of a mush of rock hard buns that had been boiled and saturated in water until they were soft and edible and could be mashed like potatoes. It was an interesting experience and quite tasty but I would not want a steady diet of it. I rather like diversity in my food, as do my colleagues, so the next night we enjoyed a nice medium rare fillet mignon coated with fresh ground pepper, prepared in a brandy and wine sauce and served with baked potato, butter, fresh chives and a side dish of brightly colored and flavorful fresh steamed vegetables. The latter meal was only available thanks to the immense background technology that brought all of the products together from various parts of the globe and placed them on the shelves of the local supermarket – the French, bless them, have taught us how to bring cooking to a high art form.

Thanks to wealth, abundance, surplus and not having to take care for tomorrow some groups in our society have had sufficient free time over the years and have managed to experiment and developed all the necessary technology to bring all of these goods together in one place and make them available to us. The technological requirement list for our steak dinner was immense; meat production, agriculture, fermenting, distilling, planting, harvesting, grinding, transportation, refrigeration, experimenting with flavors, writing the recipe, food preparation and cooking. Just for good measure we cleaned up the dishes with soap, rinsed them off and dried them so they would not form a suitable environment for organisms that might make us ill in the future. Illness is not good at sea; particularly when you are with friends. Friendships can get strained due to bad hygene.

These options are rarely available to those who live at the margins of abundance and from time to time face serious want in their lives. If the day is not assured there is little time available to dedicate to experimentation and garnering surplus so new knowledge and skills are slow to be gained, accumulated and incorporated into the community's everyday lives. If one lives at the edge of need they gather and eat what they can, when they can. If they over-gather and over-hunt to satisfy the needs of individuals, the populations of the food species are reduced and eventually the human population crashes, the survivors move and expand territory, or the population develops the technology to modify nature and get a piece of turf to produce more of the desired good or service. The agrarian society inevitably results and agricultural technology develops. At least that seems to be how the human species has worked out its bargain with nature.

Technology developed to reduce need, surplus replaced scarcity, freedom from want provided time to experiment and specialize; specialization fed the fires of technological development that in turn created great abundance. Great abundance replaced sufficient abundance and needs were replaced by wants. Needs and wants have become quite confused. Few folk, if they are rich enough, have the slightest understanding of the difference between a need and a want. With wealth and surplus, permanent settlements made more sense and planting, tending and promoting the well being of those creatures that we value economically was a more desirable option than hunting and gathering.

Such was the brilliance of the Neolithic Revolution. With the domestication of plants and animals, need was more readily met, and all forms of enquiry and specialization developed. Weaponry, architecture, engineering, chemistry, astronomy, art, music, dance, religion and transportation to name only a few were the result of surplus; of wealth. In turn codes of conduct, laws of property, specialized medicine, city planning and political structures developed. Lineage, social standing and status became more elaborate and since wealth can be passed from generation to generation if value can be captured and agreed upon by society at large the concept of money as representation of stored wealth became a possibility. Not only did money afford the storage of wealth, by common consensus money could be exchanged for some necessary good or service. It is easier to exchange a few light coins for a bag of flour than packing a heavy sack of grain from one place to another and bartering it. With wealth, greed was not far behind and with greed wealth became an end and not a means. Some got a huge slice of the wealth and others were shut out of the abundance and confined by poverty and starvation, faced an early death.

As the collection of books we humans call the Bible notes in 1st Timothy, chapter 4, verse 10: "The love of money is the root of all evil." Since the Bible was written, many humans have been attempting to reconcile that statement. War and strife have resulted, as have poverty, starvation, corruption, intrigue, violence and conspicuous consumption. The money itself wasn't the problem as the biblical verse is often misquoted as saying; it was the love of money that gets us – money as an end and not a means to an end.

Humans have experimented with various political and legal systems to ensure that each individual gets a piece of the economic pie and we haven't yet found the perfect system; in fact we are a long way from it but in some areas of the globe where technology has produced great abundance most people get enough food because starvation is neither considered acceptable nor desirable. Civil and fair society is considered important. In other areas of the globe, wealth gets highly concentrated in the hands of the few and starvation and want are the lot of the masses. Starvation and want reduce people to the level of meeting basic needs and if their basic needs are not met then they take what they can when and where they can get it – the ability of the environment to provide in the future is not particularly important if you have to step away from the game table today.

Excessive greed and excessive need among humans have always ended in land management practices that changed the environment such that it ceased to provide what we needed or desired. Overuse almost always results and the overusing species pays the consequences; we call that bad environmental management and it does not bode well for the future of the individuals who rely on the poorly managed land or for most other species struggling to capture a share of resources in the same area. Of course there are exceptions and some species benefit from poor human environments; the dandelion and the cockroach, along with others having similar colonizing or survival skills and a broad adaptability, come to mind.

We, particularly those of us in the developed countries, with full bellies might be well advised to judge less, visit societies in situations of want and examine them while remembering that we were given two eyes and one mouth and both organs should be used in regular proportion- both philosophy and strident moral rectitude are the products of a full belly. One cannot blame those in need of food for putting the acquisition of food first on their list of priorities! With regard to conservation of nature efforts and the protection of things and places of great diversity and beauty one need not travel far in their own country or even abroad to conclude that rich people; those who do not have to worry about meeting their basic needs and have considerable surplus are responsible for creating more parks, setting aside more areas of natural beauty, donating to causes they consider important than are poor people who must focus on the immediate to survive. Self-actualization in the sense of Maslov and his hierarchy of needs is pretty much a pursuit of the wealthy and not those in want. The wealthy struggle less with nature but are more concerned with it than the poor.

If we truly wish to see more areas of diverse and protected nature then the quickest way to achieving this end is by promoting education and creating wealth. Nature conservation, like high art and design, comes from excess. Titan did not paint for the poor and much of the world's great music was written in the courts of the rich. Regrettably, the rich often get rich on the backs of the poor for wealth is not evenly distributed in society. A large middle class is important for conservation. If we could make everyone wealthy it would be even better but the problem is that the wealth game, beyond meeting basic needs, is a game of relativity. If you meet more than your daily basic needs you are wealthy for the day is assured and you have a surplus, perhaps in fat, to tide you over till tomorrows chase. It seems to me that many wealthy people consider themselves poor for others have more. If one can be happy with what they have they are rich; if they aren't happy they are poor. The notions of wealth and poverty are problematic and beyond the day and meeting present basic needs the boundary is hard to draw.

Let's move on and look at how humans manage native ecosystems and the benefits and costs of good and bad management. Of course good and bad are human judgments. Weeds, and many other quite wonderful species that are not enough of a problem to get the label weed, are quite happy when humans mess up and lose out in the resource management game. At least they benefit until the disturbed system changes enough so that it is no longer suitable for their need.

Let's look at range management.

On the Management of Native Pastures

Management of native pastures for the production of red meat is an ancient resource management activity probably dating to the dawns of human transition from Paleolithic to Neolithic periods. It requires little technology to implement and has changed little between when it first appeared and its present incarnation.

Animals are tended and moved from one area to another and guided to grass and other plants that are palatable, predators are kept away and the animals are slaughtered or taken to market when they are judged to be at a suitable table product stage. Some animals are kept for breeding stock to replenish those that are removed. A balance is sought among the available forage resource, the number of animals having access to that forage and the number taken for the pot. It is a management system whose objective is sustained yield; net growth of the plants and animals on the range cannot exceed the sum of harvest plus losses – otherwise the production system fails.

In range management the rancher works closely with nature in a direct and active way to ensure the desired resource balance is maintained and must monitor the range and the animals to also ensure the range gets neither underused nor overused. Failure of the system to continually produce red meat reflects failure of the rancher to maintain a sustainable balance; we call that bad management because it is bad for the well being of the rancher and for the people who rely on the ranch products for food.

In ranching operations the native ecosystems are changed both directly and indirectly. A favored animal (usually domesticated; cattle, sheep, goats, llamas and so on) is brought into the native system and is given a favored spot and favored access to the pasture resource. Undesirable native animals who like the same forage, or who might be predators of the favored animal, are excluded or eliminated from the area. Food chains are generally shortened in the attempt to focus the accumulation of the forage resource on the body of the desired animal and keep it away from those that are undesirable. In theory the principal goals and objectives of ranching are exceedingly simple and easy to understand; in practice they are exceedingly complex and difficult to implement well. Let's look at what happens to our native pasture ecosystems when we place too many cattle on them.

When cattle are first placed on a native range they actively pursue and eat those species that they prefer; they choose the desirable plants over the less desirable plants. The desirable plants are selected and the undesirable plants are left alone; the odd undesirable one is taken inadvertently. The desirable plants are injured by the grazing action and have to expend energy to repair the damage they incur. Just as they return to good condition and start to produce forage, the cattle come along and graze them again and the process begins all over. Repeated grazing tends to result not just in damage to the above ground parts of the desirable plant; the roots below usually suffer because there is insufficient energy to dedicate to root growth to replace the root death. Roots are always dying and must be replaced and shorter root systems make the plant less able to acquire the necessary water and nutrients to support the above ground parts of the plant; it becomes less drought tolerant and less able to compete and maintain its abundance in the native system. At this point it is important to note that many native plants have adapted over time to grazing by herbivores and many benefit from a certain degree of grazing.

Too much grazing however, places the individuals of preferred plant species at a disadvantage in the ecosystem and they are unable to maintain their previous abundance. In ranching you can't take it all. The plant must be left some resources to carry it over the rough spots and meet its basic needs. These plants decrease and play a lesser role in the system. Other plants that have not experienced similar grazing pressure because they have not been grazed, and are in prime condition, are better able to capture the resources that had previously been held by the preferred species. They have a competitive advantage and they are able to increase in abundance in accord with the resources made available by the departure of the preferred species.

Many of these increaser plants as they are called, though not preferred by the cattle, are still quite palatable and full of nutrients. They are quite capable of sustaining the cattle and consequently as the cattle become hungry they begin to eat the less preferred species and settle for those species that are lower down on their preference list. The former species that had experienced no grazing pressure begin to be grazed and selected by the cattle. Like their preferred predecessors they fall prey to the cattle and have to dedicate energy to repairing grazing damage and they are put at a competitive disadvantage and can no longer hold their former abundance in the native ecosystem. With continued grazing pressure they begin to decrease in abundance and in turn are replaced by a group of plants which the cattle don't like because they don't taste nice, make the cattle ill or simply don't have enough nutrients to support the cattle.

These species, undesirable as far as the cattle are concerned at least, begin to increase in abundance and invade and take over the resources left by the earlier decreaser and increaser species. Many, though not nearly all, of these invader species are not present in the original native grassland; their seeds are blown in by the wind, carried in by animals or other means of transport. They are a part of the overall biotic potential of the ranching area. They are commonly referred to as rangeland weeds to signify that we humans don't value them; they don't work to our benefit. Seldom do rangeland weeds occupy all the available space because many of the needed resources have been lost from the pasture due to erosion and much of the land is no longer suitable for plants in its present state. If the cattle get hungry enough they will try to exist on them but over the long run the cattle will lose out and decrease in abundance through death or removal. Without food, animals die and their numbers are reduced.

We'll leave our ranching and ecosystem change story here for a moment and switch over to a human cocktail party and see if there is a parallel between the human activities at this event and the cattle on our range. Our cocktail party is an elaborate one and the tables are laden with all manners and kind of wines; rare fine French ones, lesser French ones, and fine and poor wines from many countries. The tables are laid out with an abundance of food; caviar, lobster, shrimp, smoked salmon, Beef Wellington, pickled herring, chicken, pork, hamburgers and other meat products in all types of sauces and prepared in all forms and with all kinds of appearances and tastes. There are also elaborate desserts and not so elaborate desserts; buns, exotic and not so exotic breads and crackers. There are cheeses of varying quality and scarcity and there are bowls of various nuts; macadamia, hazelnut, pecan, almonds, peanuts. Some of the nuts are segregated into separate bowls and some are all mixed together. Vegetables, vegetable dips and fruits of untold different types are laid out for the pleasure of the guests. It is an opulent gathering and the hosts have given much thought to the pleasure of their guests.

The hour arrives and the guests gather among this abundance of many specie of food. They have free and unconstrained access to whatever their heart, palate and mind lead them to pursue. Let's examine one possible scenario of what might happen at such a gathering. The scenario and the selection would vary according to the type of guest. Those of you who have attended or hosted such a gathering will probably recognize and agree with the scenario I'm going to paint. The guests at our party, at least most of them, have highly sophisticated palates and they are given to the exotic and the rare for whatever their reasons: status, honest preference, appearances, curiosity, opportunity, proximity. The possible explanations for particular selections are unfathomable because humans are complex social animals and given to all manner of motive for their actions. An all knowing psychiatrist or sociologist would have a field day in such an environment. Even our experts would have a hard time sorting out the simplest connections because they, if they were guests, would be influenced by their own preferences and emotions. Such is the difficulty with Participatory Action Research; the findings are biased by the researcher involvement. Regardless, let's move on. We won't worry about selection motives, just the disposition of the food.

First of all, our guests would likely select the fine French wines and continue to drink those in preference to the others. Then, depending on their Opemian knowledge, they would move to the finer wines of the different countries and the relative abundance of the various specie of wine would decrease according to the general preferences of the guests. Eventually they would, as their senses became impaired or their choices limited, get down to the 'plonk.' The less sophisticated wine drinkers would have had a field day and all but the most tenacious would experience no shortage of drink until the wine resource was nearly depleted. Competition gets fiercer as a desired resource is depleted.

At the food tables, the lobster, shrimp, Beef Wellington, and such like, would go first; then the chicken on a stick and eventually the hamburgers. The individual bowls of macadamia, hazel, and almond nuts would disappear first; the peanuts would generally not be selected. Peanuts are common and cheap. When individual nut bowls were emptied, the mixed nut bowls would become targets. Those guests who were unwatched would probably carefully select the favored types in the mix and the relative abundance of peanuts would increase. The occasional undesirable peanut might be picked up inadvertently because manipulative dexterity is not perfect in humans. If other guests were watching and the nut predators did not want to appear too blatantly greedy to other guests they would probably scan the bowl and reach in for their catch in that area most likely to yield the greatest number of the preferred nuts. Similar selection processes would go on with the cheeses, the desserts, the fruits, and the vegetables with their various dips. As the choice food species were depleted those guests that had enough of the food, drink and conversation would move on because it was their choice to do so. There would be fewer guests for the remaining goods but the selection process would continue and with fewer guests the rate of change in abundance of the different food types would slow down. As time passed, the prime choices would gradually diminish and only the least desirable foods would remain. Those guests that selectively chose, because of their perhaps baser tastes, the least desirable foods would have found themselves with an abundance of their preferred foods throughout the entire party. They would have experienced no competition for their preference because they used a resource few others wanted. Some food would remain untouched because nobody liked it and the tables would have much more tablecloth showing among the places unoccupied by the remaining undesirable food. The party would end and not be repeated until the host's bank account had been replenished and that would take time and effort for it is illegal to print money. Only the wealthy who had much surplus wealth accumulated and stored could afford an encore immediately: parties can be boring and one soon gets their fill. If the social occasion required or it was in the host's self-interest to have another party they might be inclined to do it again. Possibilities of future gain can persuade individuals to do things they don't particularly want to do, or even enjoy. Obligation and future benefit are powerful motivators for individual action.

As it is with our cocktail party so it is in general with the cattle and their managers on native pasture. Choice items go first, less preferable next and so on down to the unpalatable and inedible which remain because they are survivors no one wants or can use. There is one particularly significant difference however; at our party the guests had the freedom to leave whenever they had enough of the situation, their hosts did not restrain them, they could move on and pursue other activities and resources. Range managers, the hosts, restrain their guests, the cattle. Freedom to leave is not an option for the cattle; they must get by with the kind and quantity of resources available. They are captive and must get by on generosity and the wise management of their host. They don't have to worry about where their food comes from, their health, or their safety from prey – at least until the party is over. That is the contract, the agreement, one-sided though it may be. Eventually the guest must repay the generosity of the host. Some hosts throw better parties than others; they ensure there is the right amount and kind of food throughout the party. Other hosts are not quite so good at estimating, guestimating or guessing the right quantities necessary for their number of guests. Sometimes there is too much food, other times too little; good estimating requires that the host have a lot of knowledge to make things just right for all parties to the agreement.

We'll pick up our range management story where we left off and see what happens after the cattle numbers are reduced because the land that supported them is no longer able to do so. The host had not thrown a good party, there was insufficient water, not enough salt, the food was insufficient, and the bank account of the land was so overdrawn and eroded that it could not produce more food in short order. Such is the nature of severely overgrazed range.

As the numbers of cattle are reduced, there are fewer cattle looking for the scant amount of food available and they have to work hard for every bit they can get. It is hidden in the places where they do not usually go; the tops of hills with steep sides, exposed and uncomfortable places, in the cracks of rocks, among the shrubs or in the dense and impenetrable forest. A lot of cattle die in the process and the numbers get balanced. Some of the cattle are able to eke out a marginal existence; no young are born because sexual activity and its product take a lot of energy. Let's assume that our poor manager went broke, sold the few remaining cattle and moved on to greener pastures and the land was left alone.

With no cattle, the few remaining plants grow healthier, produce seed and the seed is dispersed, germinates and slowly fills in the suitable or nearby spaces with new plants. The new plants grow, produce seed, the seed is dispersed.... The depleted land gradually becomes greener.

The species that colonize are the ones that are adapted to get by on the depleted resources offered by the land. They gradually cover the land with organic matter, their leaves intercept the raindrops, erosion is reduced, the soil remains moister and the environment changes. The plant species change in response to the changing environment and as long as no new cattle are placed on the land the range condition gradually improves. Slowly at first but the greener and more diverse it gets the more rapidly the land begins to improve. Eventually the land will be covered with vegetation that resembles what was there before the cattle were introduced. Resembles but is not identical. Some new species may be there and some of the original species may not; it depends on how long the land is left alone and what the biotic potential of the area is. If seeds can't reach the area the species will not be represented in the species composition list.

The land would now be able to support cattle again if so taxed. As soon as some one discovers this rich land with its abundant forage and decides they can raise cattle ... perhaps this time....

Let's, in point form, quickly review what happened to our overburdened ranch land and the animals as a result of putting more animals on the land than the land could support on an ongoing basis. We will start with the plants and then examine what happened to the animals.

Sequence of changes to the ecosystems on the native pasture:

- The vegetation is closely grazed by the cattle.

- The plants most palatable to the particular kind of livestock are grazed repeatedly in convenient areas first.

- The most palatable plants are replaced gradually by less desirable plants until there are no desirable plants left for the particular kind of livestock.

- Fewer plants cover the ground and mineral soil is exposed.

- With overgrazing the mulch cover of organic matter is reduced.

- The microclimate becomes drier and more severe.

- The soil is trampled by the animals and becomes more compact.

- Water infiltration rates are reduced.

- Runoff and erosion are increased.

- A man-made drought is produced.

- Few plants cover the ground and much of it is bare except in small, protected areas. Rarely do all edible plants get eliminated; the few remaining are confined to out of the way places.

- The rate of energy flow and the operation of the biogeochemical cycles is disrupted.

Sequence of effects of the induced regression on the grazing animals:

- As grazing pressure increases the competition for forage among the grazing animals increases.

- The forage intake by each animal decreases as the forage resource is reduced.

- As the forage intake decreases, the grazing time increases and more energy is expended in the grazing process.

- As less nutritious or poisonous plants begin to dominate the pasture the nutritional quality of the animal diet decreases. Some death may result.

- Animal health and condition are reduced.

- The interval between birth and the first estrus cycle increases.

- The number of services per conception increases.

• The number of ova shed at each estrus cycle is reduced.

• The reproductive rate is decreased.

• The birth weight, growth rate, weaning weight, weaning age and weaning condition of the offspring is reduced.

• Once weaned, the growth rate is slower and health of the young animals is lower.

• Meat and cattle production is greatly decreased.

While the previous sequence is a broad-brush summary of the events it is generally correct. Slight variations will occur as a careful range manager continually watches all of the above, and more, for signs that management problems may be developing. The range manager will take corrective management actions in a timely manner to prevent the range from regressing so far that animal production is no longer economic or sustainable.

It is in the range manager's interest, or rancher's if they are one and the same, from a social and economic perspective to ensure the health of their range and the optimal production of the good they derive from the land. It pays to be a good host to your guests for if you are, the guests will treat you well; if you are not, you don't understand the principle of self-interest. You will lose from reduced production and reduced value of your land if you decide to sell. The value of ranchland is directly related to its ability to produce – unless you happen to be in a location where the land has greater economic value for other purposes such as urban development.

How then does one achieve the appropriate balance between cattle and forage? First we must gain some general understanding of how things work and we must translate this understanding into reasonably understandable concepts. We are reduced to concepts and must go back to the principles of classification; we must simplify so we can comprehend and from this conceptual comprehension we must intervene. We must model our system and test our theory based upon empirical evidence. We must experiment and when we fail we must correct and tune our knowledge and our actions until we get the right management regime. We must use our coarse toothed saw of concept to precisely cut the fine plank of continuous nature to suit our needs. It is tricky business and there is as much art and experience in the craft as there is science.

Experience is critical; knowledge of our particular piece of land, our particular animals, the plants of our range. We need knowledge of the specific local climate, weather, geology, soils, water, topography, range species and how these respond to the various changes in nature over time. We need to know our cattle. It is a tricky complex business and most thoughtful ranchers and managers know this; they feel it 'in their very bones' for much knowledge is based on that indescribable sixth sense that is tuned by experience. They have a general rule of thumb that works, at least until the conditions change. Wisdom is more accurate than scientific fact for it incorporates experience; science provides precision while wisdom and knowledge provide accuracy. Accuracy is more important than precision in the resource management game but good accuracy is hard to obtain without good scientific backing. Judgment of the expert is important. A student of mine many years ago, when he was answering one of my examination questions, hit the nail squarely on the head when he wrote, ". . . the biggest problem confronting the range manager is learning how to outsmart a cow." We will now have a look at range and the land management equilibrium.

As our range changed, due to heavy cattle grazing, from a highly productive meat producer to a site unsuitable to support cattle these changes were numerous and continuous. There were no distinct phases in the continuum of plant production, system change and animal consumption. One, however, can't manage a continuum for interventions must take place at a specific location so we must break our continuum up into bite-sized comprehensible units. We must classify, somehow, what we want to manage.

Let's begin by acknowledging that at the beginning of our ranching endeavor when the forage was tall and lush that it was excellent at supporting the cows. The cows ate lots, gained weight well, were healthy, happy, productive and reproductive. Our operation produced lots of meat and we made lots of money to meet our needs. Next let's acknowledge, if we are not too stressed from financial worry, that when our range was depleted and could no longer support the cows, the land was in poor condition. Further, let's acknowledge it was no accident that the range changed so radically. If we are honest, we might acknowledge our own culpability in the action; admission of error is often cleansing and can bring wisdom with self-reflection. Let's also acknowledge that we don't really want either condition to prevail. Having no cattle on the range produces neither product nor money; too many cattle on it likewise produces neither product nor money. There must be some stage in between where a balance can be achieved. We'll refine our nominal classification and, as humans are inclined to, we'll stick in a couple of intermediary states of the ecosystem's ability to produce meat consistent with the availability of edible forage plants which are needed to support our cows; let's call these intermediate states good and fair.

We now have a classification of range condition as follows: Excellent, Good, Fair and Poor; a value-based classification system that reflects the ability of the land to provide what we want. These are broad categories and include large slices of our continuous state. There are upper medium and lower ranges of each category. What they are called and how many categories one chooses is not particularly relevant as long as it is precise enough to make the judgment required for our management task – balancing cattle numbers with resource availability on an ongoing basis. Excellent range is similar to what was on our native grassland before cattle arrived; it was abundant with those plants that we called decreasers. Since it is hardly used there is no evidence of use-induced erosion. If we want to keep these plants around and our range in excellent condition we will probably not be able to put many cows on it and we won't produce much meat. Unless we are wealthy and raising cows for pleasure and not profit, or to enjoy the pleasures of native grassland while demonstrating 'to the tax man that we are serious in our efforts to make a profit,' we are probably under-using our range. It could support more for a long period. Good range condition we will define as a balance where the forage plants have a compositional mixture of decreaser and increaser species because both groups are palatable and useful to the cows, there is no evidence of use-induced erosion and lots of red meat will be produced. Fair range, we will say, has more increaser species, few decreaser species, some erosion and quite a lot of invader species. Poor range has predominantly invader species, lots of erosion and few species of any use to the cattle.

We have produced a classification that allows us to judge the condition of our range, the composition of the plants and the condition of the soil upon which the plants develop. We can easily elaborate on the details of the individual characteristics, soils, plants and their properties to help us in our decisions. Brilliant as this may be it is not enough because range condition simply provides a snapshot, a judgment of the state of the range at a particular moment in time. It does not tell us whether our range is getting worse or better and somehow we have to place the number of cows into the picture. We need to know the trend of the range whether it is regressing toward disaster or moving towards under-use. To do that we can look at details of the soil and vegetation changes over a period of time and consider such issues as: plant reproduction, presence of different age classes, plant cover, location of plants, plant health, soil erosion type and whether it is active or being re-colonized by plants, what the litter cover is and a host of little telltale signs that allow us to judge the direction of change. By an overall analysis it is possible to get a good idea of the direction for any particular condition and thus one is in a position to decide whether the cattle stocking rate is balanced or not.

The critical problem in range management is selecting the desired management equilibrium; that position on the range condition scale one wishes to maintain for their range. Once that is selected then one resorts to monitoring range trend and makes adjustments and taking management actions to maintain the desired position indefinitely; the stocking rate must balance the plant resources available. Usually that desired management equilibrium is somewhere between the middle and lower part of good. Some managers choose poorer conditions and some better depending on the specifics of the range and the amount of risk they are prepared to take for a given reward. If you manage at the upper end of fair and there is a drought period you will have no reserve grass and will either have to reduce your herd size, buy supplementary food that is usually more expensive during periods of scarcity or suffer rapid decline in the condition of your range; any number of nature's vagaries can strike at any time and cutting the management equilibrium too fine is risky business. Erring on the upper end of the good range condition is much less risky and one has more choices in uncertain conditions. In general prudence and caution are desirable but if one is overly cautious the rewards are greatly reduced. It is a tough decision but an important one and it has economic implications.

If all the ranchers 'play it tight' on the risk scale and a disaster hits they may well have to sell their stock in a glutted market and will thus get a lower price. If you manage yours on the upper end and you can 'hang on' until the glut disappears and a scarcity prevails; you will get a better price. It is quite common for many ranchers to have somewhere on their ranch what is called an emergency pasture that can be pressed into service in times of need. These pastures can be heavy forage producers if necessary with the aid of irrigation, fertilizers or other more intensive management techniques. Among other options for reducing risk, one can stockpile or purchase and store excess forage when prices are low due to climatic conditions that produced a forage glut. These are both types of hedging but one may also hedge in an economic sense by agreeing to sell so much meat at a future price well in advance of the delivery date. Hedging of any type, I must note, is also fraught with risks for the simple reason that environmental change and events can overtake us – we never know what tomorrow brings! As the saying goes, "Yesterday is a cancelled cheque, today is cash and tomorrow is a promissory note."

When judging and managing native pasture it is important to select areas that represent the average condition because cattle don't use range evenly. On depleted range it is possible to find areas where the cattle have not gone that are in much better condition, and on good range it is also possible to find areas that are overused. Cattle tend to overuse areas around water, on gentle slopes and in valley bottoms; they underuse the tops of hills, steep slopes and locations a long way from water and salt.

If ranchers can secure even forage use throughout the ranch they can increase meat production without decreasing range condition. Although it is possible, it is not easy. Cattle require water and salt to effectively use their food and consequently one can supply both of these in different areas of the range. It is difficult and expensive to develop water sources however salt usually comes in easy to handle blocks that can be set out in different areas. Water and salt close together will result in severe overgrazing as the cattle have all their needs delivered to them locally; they will survive but they won't get fat. I have visited ranches where the two are side by side and the undesirable effects on both the animals and the range are easy to see. If, however, the manager actively moves the salt blocks to the undesirable and underused places the cattle will eventually find them and as they move from water to salt they will graze the hard to reach places. It is easy to keep tricking the cattle by moving the salt blocks but it does take monitoring and effort.

There are many other techniques for fooling the cows, increasing production and securing more even use of the range and there is only one more group that I want to briefly allude to; special pasture systems. If managers feel inclined, can afford it and the land is suitable they can divide the ranch into a number of different pastures and use some intensely for different and shorter periods. They can rotate cattle among the pastures at different seasons in order to allow the plants a chance to rest, recover or set seed while the one pasture gets heavily grazed for a short time. Short intense grazing generally does less harm to the plants than longer periods of less intense use. By placing more cattle in a smaller area, all areas tend to get more evenly used and the chance of impregnation of cows in estrus increases because of proximity. There are many different arrangements and rotation schemes for special pasture management; alternate grazing, rotational grazing, rest-rotational grazing. The good range manager knows of and considers all of these options and if desirable, works toward implementing them.

Well-managed native pastures are wonderfully effective means of using a resource to produce a good while at the same time preserving native species diversity and native ecosystems. Some of the finest examples I have seen of native grasslands with their wonderful array of plant and animal species have been on ranches. It is in the rancher's self-interest to look after the land.

To conclude this section I will go back to the importance of wealth and surplus and its role in both good grass management and good native grassland conservation. The best-managed pastures are to be found among those owners who do not have to deal with an angry bill collector. If they bought the land for a good price or inherited it, managed it well and have a low debt-to-equity ratio they can afford to be more cautious in selecting their management equilibrium. If they bought the land or cattle at a high price and must meet steep loan payments they are much more likely to manage closer to the down-side edge of prudence and take more risks with the land; when nature doesn't deliver according to human plans and hopes (quite often) dreams get shattered and ranches are repossessed. Wealth and prudence, and the absence of greed and need, are good for resource management. A colleague of mine noted correctly "Neither greed nor need seem to be sustainable in the long term."

On Extensive Forest Management

In discussing forest management, it serves my purpose to make a bit of what I acknowledge to be an artificial distinction between different types of forest management. Forest management practices include the whole spectrum of intervention levels from the capital-intensive active management of intensive plantation forestry to the much lighter-handed practice of cutting down trees and letting nature at least partially reestablish a new forest from natural regeneration. I will refer to plantation forest management as intensive and the former as extensive.

It seems to me that plantation forestry and other forms of intensive forest management have quite different goals than extensive management; the primary one being to maximize the amount of fiber produced on a particular area of land during a particular time period. The primary goal is fiber production. In this sense plantation forestry much more closely resembles intensive agriculture that produces grains and row crops only the time frame for each forest crop is measured differently; in decades rather than months or a year as is more common in intensive row crop agriculture. Of course it also has many similarities with extensive forest management because the product that is produced and harvested is similar and the methods of extraction and scale of equipment are similar. The tree characteristics of interest are also more or less the same; growth rate, diameter, height, volume of usable wood, wood quality, wood properties and uses, disease considerations, stocking density, harvestable age and so on.

I also consider intensive forestry akin to intensive agriculture because no serious consideration is given to other forest species or other forest resources; in fact most other forest species are detrimental to the goals of plantation forestry if they acquire resources that take away from the maximum production of fiber. The goal is to get as much as possible of the light, nutrient, space and water resources directed to the production of usable wood. Plantation forestry is incredibly effective at achieving this human goal. Far more fiber is produced per unit area of land with intensive forestry than with extensive forestry. The ecosystem is greatly simplified, the product is known and the success relatively easily measured.

Extensive forestry is much more akin to range management because it attempts to manage complex native ecosystems with numerous plant and animal species in complex food webs and it attempts to find a balance that will ensure continued output of the desired product over the long term. In extensive forestry, unlike range management, the forest is expected to produce multiple resources for humans; not just trees, but recreation, water, forage, conservation, species preservation, habitat for all manner of valued species from ones that have huge home ranges to ones that only use the forest for some particular part of the year or the life cycle. It is expected to supply tourism benefits, aesthetic, social, psychological, wilderness and of course economic benefits and it must also allow for the exploration and production of minerals, petroleum or any of the non-renewable resources that might be present. It must make accommodations for transportation corridors, parks, settlements or any other uses that a society desires of the forest area. The production of fiber from trees is simply one of the resources among many that extensive forest management is expected to provide society.

Rather than examine the details of forestry practice which are complex and beyond the scope of this book, it is my intention to look into some of the problems that the forest planner must consider in the production of trees from forests occurring on extensive areas of public land; land that is held and managed by a government for the overall good of the society. The heart of what I want to explore is the problem faced by the resource managers, government officials thus charged in this case, in the attempt to balance all of the forest uses while still producing an economic crop of trees to serve the social and economic good of all who have a call on the forest. This crop of trees, I will presume, is guided by the principle of sustained yield whereby the forest must continue to produce for the long term. Net harvest, over the long term cannot exceed net growth plus losses and land alienations. Expressed as an equation:

Net harvest = net growth - natural losses - land alienations
+ (changes in land-use goals that influence net harvest).

The problem is an exceedingly difficult one for it is a problem of resource use optimization among competing resource users, land-use interests, and social goals and aspirations. It is not a problem of product maximization where the production of one resource is the primary goal and all considerations can be predicated to achieving that one goal. Mathematics informs us, quite correctly, that one cannot maximize two variables simultaneously so consequently in extensive forest management those charged with the management task must find a balance among all desired products and services.

In a democratic market society rarely does the government get directly involved in the actual forest harvesting operation. Its role is predominantly that of setting policy, allocating the forest resource fairly to those that will extract it, ensuring that the renewable resource is adequately replenished, approving plans for extraction, ensuring that forest policy is implemented in accord with the prevailing sense of the social good and finding the optimal balance among all interests and uses. The government managers are faced with an Integrated Resource Management (IRM) exercise and they must develop Integrated Resource Plans (IRP) to guide the forest management. It is not my intent to judge the appropriateness, success or failure of the government managers in their efforts at forest management for in democracies the people judge government actions at the polling booths.

Let's examine the requirements for managing an area of forest for fiber production on a sustained yield basis. There are a number of basic requirements:

- A permanent land base.

- An adequate inventory of the existing forest resources in the management area and a knowledge of the productivity of each of the various forest sites over the whole property.

- Adequate protection of the forest resource from insects, disease, fire and other natural or human caused losses to ensure they do not exceed an allowable maximum.

- Regulation of the amount of fiber removed, the cut, such that removed resource during a selected time interval does not exceed the net growth minus losses during that time period.

- Quick renewal of forest growth after harvesting or other losses.

The requirements are simple; their application is perhaps as complex as any renewable resource management problem challenging humans for it must deal with the vagaries of changing nature over a long period into the future, perhaps hundreds of years or more. It relies on a good understanding of how much usable wood presently exists and will be produced over the whole management area on a site specific basis; it requires some sense of the wood product desired a long distance into the future and an understanding of current and potential human goals and aspirations for that piece of land over a similarly long time span. Time is an enormous problem. The farther into the future one must look and plan for the more uncertainties and unpredictable events will or are likely to beset the forest management area. Any one of the above can and in all likelihood will change. The changes in many areas will not be small. Let's build a simple model and see if we can elucidate some of the problems and how they might be dealt with.

To begin with let's assume that we have a piece of public forest land that is one thousand square kilometers in extent, has never been logged or cleared and we want to manage this area over the long term on a sustained yield basis so that it produces a continuous output of wood on a yearly basis into the future; perhaps in theory forever. We will call it a Public Sustained Yield Unit or PSYU for indeed it is land owned by the public and we intend to manage it on a sustained yield basis. It is desirable to have large areas for a PSYU because a large forest fire or insect outbreak on a small area can have a huge impact on the amount of wood available over time; it is less significant on a larger area. Small management areas are too risky and more or less destined to failure. We will also assume that we want to produce a predictable and regular volume of wood each year because it is socially and politically desirable to provide regular and continuous employment for the employees working in the forest. Furthermore without continuous and predictable production it would be impossible to attract the investment capital necessary to build infrastructure and supply equipment; forestry is a highly capital intensive business.

Next, for the purposes of our exceedingly simple model let us assume that all areas of our PSYU are uniformly productive. This assumption of course is patently incorrect for nowhere on such a large area of land is the soil, topography, climate, moisture holding properties or nutrients ever uniform. Furthermore the topography on such a large area will vary immensely. There will be areas of sandy or clayey soil; tops of slopes with less moisture and bottom of slopes with more moisture and nutrients which have been brought in from groundwater movement; variations in local climate caused by the topography and other climatic influences and over a long period there will be considerable variation in the regional climate; there will be variations in the geology of the areas which will affect nutrient availability. These are only a few of the vast sources of variation in forest productivity that will occur on our area; the possibilities for productivity variation are endless. Somehow we must determine the extent of this variation as accurately as possible and make provision for those unforeseen events that might occur in the future but are totally unpredictable; fire, insect outbreak, disease or extensive blowdown.

It is possible to determine the productivity of our whole site with a careful forest site productivity classification and mapping program before we begin to harvest but it is time consuming, expensive and at best approximate. Such an inventory and classification is essential for without it we cannot predict how much wood will be produced and consequently we cannot manage the property on a sustained yield basis. The more accurate the inventory and site classification is the more accurate our predictions will be. As far as unforeseen future losses to insects, diseases or fire we are reduced to an assumption, hopefully, based on past experience and analysis of past events. Let us assume it is in the order of one tenth of one percent per year; this is not unreasonable for many forests and should it look like our losses will exceed these we will commit the resources to suppress the fires or control the forest pests as best we can. Our site productivity inventory is critical because there is immense variation in the ability of land to produce wood. I have measured sites over a large area in Alberta and productive sites can produce more than ten times as much wood as poor sites. There will be some areas in our forest that produce no trees; lakes, swamps, dry grassland areas. We also need to know how productive each site is for the various tree species that we wish to grow and harvest, for each species has different resource needs and responds differently to various environmental factors. One species may grow well on one type of site and another species may grow poorly; productivity is species specific.

Based upon the findings of our research into forest productivity let us assume that we found it takes one hundred years to produce a tree that is merchantable and ready to harvest. We will call this our rotation age; the length of time it takes to grow a tree to a specified age and stage of maturity. We will manage our PSYU over a hundred-year timeframe. No more, no less and we will determine our annual cut based upon what we think will be produced during that period. This annual cut is usually referred to as the AAC or annual allowable cut. If, on average, we cut more we will not be complying with our assumptions and will probably end up with a timber shortfall at some future time; if we cut less we will be harvesting less than we could. Generally there is a bit of flexibility in the AAC to allow some sensitivity to market prices. It is a similar problem to the problem of the range manager who must balance grass and cattle production.

Our initial assumption was that our area currently had trees on it but we did not specify what species were represented, how abundant they were, what stage of maturity they were at, their age class distribution or what condition they were in. If the whole area is covered with juvenile and unusable forest then we cannot even begin production until some of it begins to be mature enough to harvest. If it is all too old then we have a problem also because everything will be ready to harvest when we start and over a long period the mature trees will be infected by insect, die or slow down in their growth for trees do not keep enlarging forever. This problem is another reason to select a large area for our PSYU; rarely over large areas does nature ever provide uniform forest age classes. To overcome our knowledge deficiency of current forest status we need to produce an inventory that documents and maps species, volume, stage of maturity, condition, location, distribution, accessibility, non-productive forest and a host of other important things. For our simple model we will assume that nature in our PSYU has been obliging and that we have the perfect distribution of forest types, age classes and other matters related to present forest condition.

We will quickly review the assumptions of our simple model: one thousand square kilometers, uniform productivity over the whole area, one hundred year rotation age, existing forest with uniform distribution of usable wood at exactly the right time, perfect distribution of types and ages of present forest, forest losses not exceeding one tenth of one per cent per year, and immediate renewal of forest growth as soon as soon as we log it. Totally unrealistic assumptions but we have discussed a good number of the problem areas associated with each and have some appreciation now of the complexity and uncertainty faced in producing trees on a sustained yield basis. If we can't make the simple model work then we really don't have a prayer in applying the management model to a real piece of nature.

Don't worry, I have set our simple model up for success: in theory we should be able to harvest 10 square kilometres (1000/100=10) of land minus one tenth of one percent or 0.1 square kilometres (10/100=0.1) allowance for forest losses each year. We should be able to harvest that each and every year for as many forest rotations as we are inclined to do. By the time we finish cutting our hundredth annual forest block, the first one will be ready to harvest again. We win! The only problem is that it is extremely easy to make things work when we set the rules and everything conforms to our rules; mathematics is perfect and 2+2 always equals 4 providing we are working in Base 10 and there are no unknowns and uncertainties. Perfect knowledge (and inaccurate theoretical models) usually generates perfect results however we never have perfect knowledge of the present state of natural things let alone perfect knowledge of what will happen in the future. Theoretical models can be useful for pedagogy and examination and in our case we have managed to get the wood production side down. What else are we missing?

We are missing a lot. In fact all we have considered in our simple model is the simplest and most predictable of things, the growing of trees, and trees are reasonably predictable in their growth over the long term. Take any old tree, cut it down and examine the annual rings and you will find that the growth is fairly predictable over the long term providing the site has reasonably deep soils with good moisture holding capacity (trees growing on thin soiled sites show much more annual growth ring variation as they are much more sensitive to variation in precipitation; thin soils do not store water and buffer moisture variation). Certainly there will be growth variation from year to year; there will be years when the tree got fewer resources and produced smaller annual rings and lower height growth and other years, or periods of years when the annual growth was much larger. Growth variation always occurs and the variation can be caused by resource availability changes within the stand (autogenic factors such as crowding or shading) or outside the stand (allogenic factors such as climatic and weather patterns).

Let me take a small detour in our planning exercise for a moment to discuss tree growth. Years ago, when I was doing some research just a few meters from the Arctic Ocean I noticed a small White Spruce tree growing prostrate to the ground. It was 2.5 cm diameter at its base and a bit more than a meter long and I was intrigued to find this creature so far north and curious to know how old it was. I cut it down, with great difficulty for the wood was exceedingly hard, and took it back to my laboratory with the intent to age it. I thought the problem would be simple; put it under the microscope and count the annual rings. I was wrong for the rings were so small and tight that they could not be counted using any of the standard dendrochronological techniques and preparations. My curiosity increased and I solicited the help of some colleagues and managed to get some time on the Scanning Electron Microscope. Even the SEM didn't make the task easy and after much difficulty, and using ring matching techniques, I concluded that it was likely that the tree started somewhere in the 1860s; it was over 100 years old. Some years the tree was so pressed for resources that it had no decipherable growth in size. Net photosynthesis was balanced by respiration; the tree used every bit of sugar that it produced simply to live. This little project was enormously instructive to me.

Back to our IRP problem. In our simple forest management model we have not considered the biggest unknown and unpredictable of all. We have not considered humans and what calls they might put on our thousand square kilometre land base over a minimum of one hundred years let alone over much longer periods; we have integrated no other resources and we have not allowed for human values. We have integrated effectively nothing. Back to the drawing board to reconsider our plan and consider what changes might take place on our IRM area.

Over one hundred years it is likely in such a large area that we will decide we need to build some roads, designate some parks, explore for and develop some oil or gas wells, construct pipelines, develop a mine, establish a wind energy farm, build a reservoir or set out a solar collector array. We might also decide to designate some ecological reserves, establish recreation areas, expand agriculture, protect breeding grounds or travel corridors for animals we value. A town might be established; perhaps an airport. The possibilities are unlimited and we certainly don't have a clue where or when these calls on our land-base might arise. A hundred years is a long time and human needs, wants and desires are unpredictable. With each development and land use change our wood production area gets squeezed and reduced because they all have high space requirements. Our ability to balance net growth with harvest plus losses becomes more difficult; difficult though not impossible. After all, with a smaller area to grow trees on, all we have to do is reduce the amount of timber we harvest each year by lowering our annual allowable cut. It should be simple because we simply have to recalculate our input/output model and adjust the AAC accordingly. Let's now examine what might happen when we reduce our cut.

When we reduce our AAC we may find that our milling and processing facilities are now over-built for the amount of wood available to process. The investors who built the mill on good faith of a continuous and regular supply of wood for a fixed period of years, and had expectations of recovery of their investment plus some allowance for profit, are unhappy. They squeeze government to maintain the wood supply, find substitute wood from an area outside the PSYU or provide financial recompense from the public purse. If the mill reduces the amount of wood it produces it will need fewer employees so work lay-offs occur; the size of the forest sector economy will be reduced which in turn affects the local shop keepers, banks, bakers, trucking firms, consulting firms and on and on down the line. When a local economy is driven primarily by the production of a single resource it is exceedingly fragile; when that resource is depleted or reduced and unemployment abounds, the personal and social tragedy is enormous. The government is placed under considerable pressure to maintain or shore up production and keep the economy going for the social and economic good of the people. They are also placed under enormous pressure to keep the resource flowing because without it the government coffers will be less well endowed and government programs will have to be reduced or debt incurred to maintain them. The electorate, locally if not at a larger scale, will become unhappy and the government will run the risk of being replaced. Ruling governments don't like this option and do their best to avoid it. If the economy is diverse and doing well in other sectors the government is more likely to be brutal if it must and reduce the AAC in the forest sector; if the economy is poor in all areas, brutal pruning is much less likely. In any economy, good or bad, politicians are not terribly inclined to take tough decisions that will make the electorate unhappy so they do whatever is possible to keep the socio-economic pot boiling. They exhaust possibilities and attempt to find an alternative to reducing the AAC if at all possible. This is usually reasonable because they are elected to serve the needs of the people and technical arguments supporting this decision can be mustered by forest managers.

In the case of our timber supply and forest resource management problem there is an alternative and the scientific and rational arguments are powerful to support the alternative. Here are some ideas.

When we first began our forest production operation we undertook a comprehensive inventory of the resources in order to calculate the allowable annual cut and undoubtedly we discovered areas of our land base that were not producing trees to their maximum capability. Those areas that were not fully stocked with trees could have more trees planted on them after logging, areas of currently non-productive forest could be altered a wee bit – swamps could be drained and planted to trees, artificial fertilizers could be applied. Forested areas covered with species that are not of current use could be site converted and replanted with desirable species or perhaps the technology will change and uses found for the undesirable species. The forest utilization standards could be changed so that there is less wastage and more product garnered. The list of alternatives is large and we have not even touched the possibilities available to us in the area of forest genetics. We could manipulate the forest genetics and plant trees that were carefully selected so they grew faster, produced a higher wood quality and were worth more on the market.

In short, our possible ecosystem manipulations are immense and we could gradually shift from extensive forest management to slightly more intense forest management; just enough to keep the original balance of wood production and maintain a balance. Our tool kit of ecosystem manipulation techniques is large and we can make sound arguments in support of maintaining our original allowable annual cut and keeping everyone reasonably happy. Our backs are not yet against the wall and we have lots of alternatives to assist us in meeting our goals. In fact we have so many possible future alternatives that we don't have any serious worries. If need be we can even change our extensive forest management policy and gradually crank the production screw tighter and eventually get down to changing the policy to allow for intensive forest management. We could eventually get down to plantation forestry and produce the same volume of wood on a fraction of the area required for extensive forest production. Intensive management of natural ecosystems is also a possibility. The only problem is that as we travel this path, the amount of uses or goods the forest provides diminishes – the path is one that leads from a forest providing diverse resources toward providing a single resource. Forest values must change as we find ourselves on the path to intensive ecosystem management.

On Intensive Ecosystem Management

Intensive ecosystem management for the production of any number of desired goods or services is always an option and it is, if we can keep the intensive system stable and producing, far more productive and capable of meeting specific human needs or wants than extensive ecosystem management. One can produce much more from a much smaller piece of land; more row crops, higher forage production, more fiber, more fish, more shrimp, more livestock, more bananas, more coffee, more berries, more caviar, more and higher quality wine – more, more, more....

Intensive ecosystem management was the magic that came out of the Neolithic revolution with the domestication and management of plants and animals. This resulted in increased production thereby allowing the storage of surplus during the good times to tide one over the bad times. Storage takes the peakiness out of natural systems and it is not an idea that is unique to humans; nature discovered the trick of storing surplus at the dawns of time. Plants produce and store sugars and carbohydrates they need to tide them over droughts, harsh winters and other times of want. Animals store fat. All these storage products and their various techniques can be converted to complex sugars that are broken down to simple sugars which in turn are processed through the Embden-Meyerhoff pathway to glucose. Glucose, in turn, goes through the Krebs cycle or the citric acid cycle to convert adenosine diphosphate to adenosine triphosphate. These in turn can release energy for our fundamental life processes. That is chemical storage to meet relative short-term physical needs. Nature has found, through genetic error, an efficient system of storage and release of energy. Humans have, through copying nature, extended the storage concept much further. We can store for years, decades, centuries ... perhaps millennia; honey stored thousands of years ago is probably consumable. Canning, drying, preserving, salting, irradiation; anything to keep the microbes at bay. The flavor may be reduced but our basic needs can be met – we can survive on food that has been stored for a long time. All we need for surplus is vision and experimentation ... and energy; lots of energy. We will leave this for the moment and come back to storage and energy at a later point. For the moment I'd like to go back to intensive ecosystem management.

Intensive ecosystem management is nothing more than manipulating natural, native ecosystems such that as much of nature's resources as possible are directed to the product we want. Once we stop our manipulations of the natural ecosystem, natural processes gradually change our manipulated site back to a complex array of ecosystems consistent with the biotic potential of an area and the physical and chemical characteristics of the environments. I can neither think nor imagine an area of the globe that would not, over a long time, erase the evidence of any changes brought about by any species; including humans. One need only reflect on the fate of human artifacts from past cultures. The evidence is sedimented, eroded, decomposed, consumed, covered and hidden with vegetation – Mohinjudarhu, Machu-pichu, civilizations in the Middle East, Africa, the Lost City in Cambodia.

How do we manipulate ecosystems? Effectively our manipulations fall into a few broad categories. We manipulate the controlling variables of climate, geology, topography, hydrology and available organisms in an area. We also manipulate the time factor it takes for nature to change. Furthermore, at a much more refined level, we can manipulate the dependent variables operating in a particular system; native organisms, livestock, vegetation, soil, decomposers, microclimate and we can manipulate our own manipulations.

We modify the climate by acts such as cloud seeding, burning, deforestation, heat and chemical emissions. The geological materials and the hydrology are altered by leveling land, terracing, erosion control, spreading water through all types of irrigation, nutrient enrichment, fertilization, dams, river flow control, drainage and groundwater recharge and discharge, additions and changes in the arrangement and sequence of types of geological materials. The biotic potential and available organisms are manipulated by species introductions, species elimination and genetic manipulation. Most of these manipulations are deliberate; some are inadvertent changes that we did not foresee.

We change the vegetation by plant control, planting, tending, spacing, thinning, harvesting, herbicides, forage management, range management, forest management, genetic manipulations and controls, disease control, rodent control and the control of other pests. We revegetate areas that have been denuded of vegetation. We actively manage the soil by tilling, plowing, scratching, scraping, topsoiling; addition of nitrogen, phosphorus and potassium and plant micronutrients. We add acids and bases to balance the pH and we change the soil structure, consistency and texture by various means. The animals are manipulated by grazing management, insecticides and other pesticides, livestock management, fisheries management, breeding, crossbreeding, artificial insemination, reproductive control, disease and health management, antibiotics, hormones. Microclimates are changed by the use of shelters, mulches, shades, buildings, canopies, barns, heat-pots. The various techniques, manipulations and equipment available to change and manipulate ecosystems are an overflowing toolbox of options and possibilities that can be selected or combined in any array of combinations and permutations over innumerable time-frame options. A short visit to the local library, university, technical school, government department or equipment and supply dealer to enquire about equipment and supply availability is truly enlightening. If you are not inclined to move from your chair you have access to a phone or the internet where you can spend as much time as you wish dialing, talking, searching, surfing and reading. All you need is curiosity, desire and motivation.

All of these manipulations in turn are guided and determined by what we want, what we know and what resources are available to us; our goals and objectives, available techniques, beliefs, fears, needs, wants, goods and services, research and of course economics. Things rarely get done unless there is a belief that they make economic sense at either the micro or macroeconomic level.

Economics, at least in market economy societies, usually is a key driver of ecosystem manipulation because most people do things for economic gain; unless of course they have an abundance of resources, or at least sufficiently more than they need for the foreseeable future. In the latter situation they may be inclined to manipulate ecosystems out of benevolence, for altruistic reasons, perceptions of social justice or morality or for fame and social admiration.

Economics at best is an uncertain science and it is driven greatly by belief and faith. Examine any business plan and it is not hard to uncover the acts of economic faith involved in estimating market potential, demand, projected income and expenses and a host of other beliefs that are put forward to convince the banker or other investors that that the risks are acceptable and economic success likely. Perhaps it is unfair to economics to even categorize it as a science; it might be better to consider it high art. Perhaps abstract art where the economic artists paint a representative picture of their perception of the future economic reality based on the past and present situations. Past economic performance is not always a reliable indicator of future performance as many business people and investors have learned at the cost of considerable pain and hardship. Markets are not rational because people, particularly when they are part of a crowd, are not rational. Recall examples of economic madness when good folk are overtaken by greed and lose their senses – the Dutch and tulips in the 17th century, the South-Sea bubble, real estate in Florida, the Internet Dot Com bubble. Not only are people not rational, some of them are downright deceitful, manipulative and dishonest.

On Intensive Carrot Cultivation

Let's make a carrot patch: a large one to feed a lot of people. Carrots – because we think people like lots of carrots or because we believe it will help their eyesight. Whatever our nefarious rational, we have made our decision. We won't worry about bunnies too much even though as a child we learned that bunnies like carrots. Besides there may be no bunnies on our land when we buy it and if there are they may not be a problem. If they are a problem we can always shoot them or eliminate them in some way. If we want a few around because we like bunnies we could always feed them carrots laced with progesterone or estrogen so they won't be quite so prolific. Perhaps there will be coyotes on our land and coyotes like bunnies. Now if the coyotes become a problem ... well, we could feed them bunnies laced with.... We've got the bunny and coyote worries under control should they arise but they won't concern us unless we get some carrots and the land to grow them on.

Off we head to the local library to learn about carrot types, their nutrients, water, space and climatic requirements. While we are at it we will see what is known about production techniques just to save us from having to experiment too much – we remember from school what Isaac Newton was purported to have observed, "I have made my discoveries by standing on the shoulders of giants." We know we are no Isaac Newton and besides we are not interested in celestial mechanics; we just want to grow carrots – lots of them. We want to invest as little as possible and turn as large a profit as possible.

Our visit to the real estate office proves fruitful. We find the ideal piece of land with perfect soil, climate and topographic conditions and an abundance of native plants and animals on it. Furthermore there is lots of cheap water available, good transportation and other infrastructure and it is near a large city with a big potential market. We have a couple of added bonuses; the price is right and it is close enough to the city that the land might go up in value just in case our carrot patch venture doesn't work. We could always become a land developer – cities usually grow and expand either because people are like rabbits or the living is easier in the city.

We solicit the help of contractors and break the sod, turn over the soil, level the land, remove the native plants and when they have gone we won't have to worry about the animals because animals need those plants. We set up an irrigation system and a reservoir to store water and supply the sprinklers just in case we are faced with a long dry spell. To control possible problematic plants we give the land a shot of herbicide – we want to start with a clean slate.

We prepare all the land with the right number and spacing of rows for the carrots and we seed all the rows with the right spacing of the right kind of carrot seed that has been genetically manipulated to provide high productivity, the right shape, the right flavor, color and the appearance according to what we think will ultimately appeal to consumers and before them, to the wholesalers and buyers.

We spend the next few months watering our land, eliminating weeds, controlling insects, fertilizing, laboring – dreaming of our bountiful harvest and the money we will make from all the happy customers. Life is good for a while until we realize the only money involved in our operation is going from our pockets into someone else's. We worry a bit but ours is a new operation, we are committed and we hang in until our carrots are ready to harvest and sale. The crop is bountiful and our carrots are wonderful; we begin to harvest and look for markets. The harvesting is easy but the markets are not as anxious for our product as we would like and besides, other carrot growers are bringing their carrots into the market at the same time and the competition is fierce. We sell enough of our carrots to cover our debt obligations for a substantially lower profit that we hoped – at least we turned a profit even though we hardly drew any salary ourselves. We still have some stored carrots to sell later when the markets improve. We survived financially but during the long winter we will do a bit more research and see if we can find a more efficient and productive way to operate things just to even out our cash flow problems.

Perhaps we could plant our carrot seed more densely and thin the rows several times over the summer. That way we could produce and sell early baby carrots. People always want fresh young carrots and they will pay a higher price. Besides, the carrots last year did not use the space, light, water or nutrients efficiently; carrots only need a lot of space as they get bigger. We will multi-harvest. Now that we think about the space, water, light and nutrients issues perhaps it would be possible to interplant green onions or lettuce between the carrot rows. Lettuce and onions grow quickly and they will be finished before the big carrots need the resources. We won't need any more space, fertilizers, or pesticides. We give some thought to growing organic vegetables but we know the quality, color, size and productivity will be reduced and the labor costs for weeding will be much higher. We also know the market for organic vegetables is small and vegetables markets are also highly price sensitive. Too risky! We vow that next year, when our cash flow is better, we will buy our fertilizer and pesticides at the end of season when prices are lower and stockpile them. After all, every penny we save in our operation goes straight into our pocket.

Next season comes and we apply what we have researched and learned over the quiet winter and we do better. Our onions, lettuce and baby carrots have smoothed out the economic peakiness a bit and our productivity is better. We have a few problems with weeds and insects but thanks to chemistry and the research behind our industry we overcome them.

We realize that if we could only deliver our crops a week or two earlier next year we could command a higher price and at the same time increase our market share. Maybe we could plant earlier and cover them with black plastic film to absorb more of the sun's heat and warm up the soil sooner. Perhaps we could switch to clear plastic once they germinate to save moisture, keep the air warm and still let in the light. We will also manage to buy our chemicals at the end of season and that should help the production economics. During the summer we noticed the high winds were desiccating the soil and we have some erosion problems. Perhaps we could plant some fruit trees around the perimeter to reduce the wind speed and erosion. We might also trim the tops off the early carrots before we sell them and leave the waste on the soil to provide a mulch cover to reduce our water loss. When the season is over we will plow them in to enrich the soil and reduce our fertilizer needs. Perhaps we could squeeze our labor costs a bit or increase the mechanization.... Change from carrots to something more exotic that will command a higher price? Perhaps we could process our own carrots and market them under a brand name? Carrot juice? Maybe we could raise bunnies; they like carrots.

The possibilities are endless in this game and experience, research and knowledge make us better and better carrot growers. In fact we know so much about carrots and love them so much that carrots are all we talk about. The friends we asked over for supper last weekend for the third time said again they couldn't make it. Perhaps we could take a night course in celestial mechanics just for interest ... who knows perhaps if we understand celestial mechanics better we will understand how the sun and the earth interrelate and we will be able to improve our carrot productivity....

A silly little example of intensive agriculture, perhaps, but one that is instructive of the type of continual refinement and improvement that characterizes the intensive production of anything; fish farming, forestry, fruit production, shrimp farms ... pretty well everything one can think of.

It is immensely productive of a desired or needed product in terms of land base use.

Intensive ecosystem management relies on manipulating natural ecosystems by simplification in the extreme in comparison to the complex array of species interactions with each other and their environment that nature, in the absence of human intervention, has found successful over the long term. It relies on an extremely high level of control and manipulation of energy and resources in order to direct energy and resources strictly toward increased production of the product or service we want. Other species that normally would occur in the area as a result of the biotic potential are kept out and deprived of the resources they need. Intensive ecosystem management is, however, still a form of natural ecosystem management for it still relies on a fixed land base and the solar energy reaching that land base. It relies on local environment, to provide appropriate temperature, water and nutrient resources to meet the needs of plants or animals being produced; these needs are almost always supplemented by further additions of nutrients and water. Intensive resource management always involves species elimination, species introduction and manipulation and system simplification.

As I said in the opening of this chapter, intensive ecosystem management is extremely efficient in producing what we want but I also placed a provision on the productivity...the trick in intensive management is to keep the intensive system stable and functioning in the direction we wish. This is particularly difficult because the system has been so simplified that it has little system redundancy remaining in it. There are few species on the land base and they all, at least the crop species, have more or less the same call on resources as each other and they all use these resources at more or less the same time. They all have the same growth form, nutrient requirements and they, because of their particular biological properties require these resources at the same time. In their maintained state these systems are not particularly adaptable to changes in nature's abundance. Furthermore they are crowded in the extreme and spaced according to what humans have found to be optimal for maximal production. If humans 'get it wrong' the production will fail and little or nothing will be produced; if they get it right they are big winners.

In ecosystems unaltered by humans and in situations involving extensive resource management, the system is much more biologically diverse and therefore robust. Some plants require their resources and go through their life cycles in the spring, others in the summer, yet others in the fall and some in the winter. The timing of resource needs is continuously changing as the days pass and the waste from one species becomes the food for another. This overlap in timing and variation in resource requirements provides the system with a high level of complexity and redundancy; it has resilience to it, it can bounce back, change and adapt to change as needed. If conditions are such that one species fails, there are other species with similar requirements that can fill in the lost role and capture the newly made resources.

Intensively managed or simple systems are a bit like the pond in my back yard in its early days; prone to peakiness and wild swings until more species joined the system and their wastes settled to the bottom and provided food for others. The wastes from these organisms combined with the non-biological part of the environment to provide storage of the required resources for the whole system, and the stored resources buffered the system against all but the extreme variations in climatic conditions.

Intensive resource production ecosystems are also a little bit like the highly specialized orchid we talked about earlier. They are so specialized that any significant change in the environment will make them vulnerable, will perturb them and throw them into a bit of a tail spin. Specialization has its benefits providing conditions remain constant and change is within acceptable limits.

Unlike the orchid growing in the forest, if the desired species fails there are no other troops to soldier on and pick up the slack. That is of course, unless the humans replant or restart the system and try again or simply walk away from the endeavor and let natural processes take over and reestablish a system that is compatible with the prevailing physical and chemical site conditions and the prevailing biotic potential of the area. Historically nature has been good at doing this; five billion or so years of history and untold numbers of extant or extinct species are testimony to nature's ability. There is much to be said for the imperfection in the elements and the genetic code.

Humans, a bit like nature, are not particularly put off by their failures. They pick up and try again. They reflect on their failures and learn from their mistakes and plan other strategies that might reduce the risks and lead to success. Perhaps, just perhaps, we can keep the problem organisms out, increase our control over the needs of the organisms we want to produce. Perhaps we can totally control the needs of the desired organisms; their light, climatic, nutrient and water requirements.

Perhaps, just perhaps, we might be able to eliminate nature's whimsy and unpredictability. Perhaps, just perhaps we could get more specialized and grow our products in barns, greenhouses, vats or ponds. Perhaps, just perhaps, we could only let those things into these facilities that we wish and only let those things out that we either choose to or must.

Perhaps, just perhaps, we could try for total environmental control, eliminate risk and further increase the productivity on a smaller piece of the earth. Our dreams, our knowledge and our technology are powerful.

On Industrial Biological Production

Systems engineering and systems control are not new to humans by a long shot. For thousands of years humans have been designing systems and controlling them. A long time ago, the first cooking pot with a lid was used with fire. This is a simple system but a controlled system nonetheless. A certain quantity of product is placed in it, a lid placed on top to prevent water and heat loss and the temperature underneath it controlled to provide the desired effect inside. When the desired effect was achieved the product was taken out: a highly controlled environment.

Other early examples include apparatus to produce wine that ferment the grapes with the assistance of desirable yeasts and keep out some of the less desirable wild strains that reduced the quality and taste. Cheese production and the baking of leavened bread rely on microorganism and fairly elaborate apparatus. Ancient arts discovered by chance, were elaborated by trial and error and empirical evidence over the years to the level of high craft. They are ancient examples of systems engineering and control but early on there was little known about the details of the theory behind them.

Today we have elaborate knowledge of how to engineer systems; static systems, dynamic systems, input/output models, process theory and various system controllers. We also have knowledge of feedback loops, systems monitoring and automatic control – all of which can be designed to afford a much higher level of accuracy and precision than human judgment is capable. These can all be combined in the appropriate way to make a highly efficient process that allows us to determine how much resource to put into the system, how to process it, what the desired product will be and how much and what quality of the product will come out the other end. The more we understand about the system the more we can refine it and increase its efficiency and the better it can serve our needs.

System control is important to humans and nature has experimented with it a lot. If we consider humans we realize they are like a complex system; in fact a series of complex systems, all of which interact at some level of harmony if the human is functioning properly according to its genetic design specifications. The respiratory system, vascular system, alimentary system, reproductive system, urinary system, immune system, digestive system, musculo-skeletal system, nervous system and so on.

The reproductive system is not essential to the individual organism but is nice. When one of these systems gets out of control, ceases to control or be controlled in accord with the total system, the human, is considered ill, disabled, disturbed. All systems must work in harmony or the system is at risk. They don't have to work in perfect harmony but they must work well enough to keep the organism alive and prevent other organisms that constitute part of the biotic potential of the human's environment from penetrating and disturbing the system. There must be a certain balance among all the systems and they must control each other interactively and they do: too cold – we sweat; too hot – we shiver; too tired – we sleep; too much waste build-up – we defecate or urinate: automatic system responses to the external and internal environment of the organism. The organism is self-regulating. As long as it is alive the organism has little or no choice over these functions.

Some functions we do have a choice over but I won't go into these as they are not particularly critical to industrial biological production systems. In fact choice is usually undesirable in such systems because choice reduces the level of system control by the system designer; unless of course the designer can constrain the choices to only those responses that are desirable. Choice is not a good word for it implies a certain degree of individual initiative which does not exist. It must do only what we want and we want to prevent the system from failing as well as we can. We don't want our industrial biological system to be as sophisticated as a human because humans don't always do what they are told. Our system must have limits and it must be constrained. It is a little tough to understand systems control in humans so let's examine two simpler systems. Humans and other living organisms are just too wonderfully complex to understand and we humans just don't have life sorted out yet although we keep trying. Let's consider the gas pedal of a car because this is an externally controlled regulator. After that we will briefly examine a thermostat that controls the heat in our house.

Through a series of levers, springs, valves, pipes and pumps the driver is able to exercise choice over the rate of gasoline delivery to the engine and the consequent power delivery to the wheels. If only a little gas is provided the engine runs slowly, a lot and it goes faster. Too much gas and it either goes too fast or the engine blows up and ceases to function. No gas and the engine stops. When we drive a car we have come to expect a perfect response. The system must do as it is designed to or we have problems. Our system however is a discretionary system; the driver exercises control over the input and the output of gasoline passed from the tank to the engine. It needs regular and careful attention by driver in order to achieve the right dynamic balance or to stop the system and make it static and non-functioning. Such control is desirable in cars but is time consuming and if one has to watch and manually maintain the balance of every system they design, a lot of time is used. In some situations the task may be boring and repetitive and humans are not good at boring and repetitive tasks; they are given to mental wandering, dreaming and failing at such tasks.

Consider the fate of a human being that had to decide when to breathe, when the blood should be pumped, when to shiver. Exercise on a continuous basis of choices like these is clearly not desirable in humans. So much time would be consumed in actively controlling these systems, they would not be able to look for food, chase prospective mates or watch out for predators. These systems must be controlled accurately and precisely and they are appropriately taken out of the arena of choice. Other systems appropriately are given a higher level of discretion in when they function. In the case of humans there is some social and health merit to having a bit of freedom to choose when to urinate or defecate. One does not want to lose friends or foul their own nest. In other systems it is desirable to have as much control as possible; locomotion, arm and hand movements, hearing, vision, eating. All systems must function correctly and with just the appropriate amount and degree of control.

Precision, accuracy and timing are all important in systems control. Let's consider the thermostat that controls temperature in a building or any other system that requires automatic temperature control. Thermostats control a variable, temperature, which the operator selects according to choice. There is a performance range and accuracy level that usually is set once according to desired performance specifications and only changed when there is a problem. We know pretty well what precision level is required for humans to be comfortable so the thermostat precision is set such that the thermostat will kick in approximately 2 degrees centigrade on either side of the desired level. Set the thermostat to 20 degrees and the furnace will come on at 18 degrees and shut off at 22 degrees. Change the setting and the automatic control range will remain the same; 2 degrees either side of the desired level. Now, human desires and the performance ability of the equipment being controlled must match; failure to match performance and needs will result in either an uncomfortable human or equipment failure; in either case the human will be unhappy. Such systems are sometimes referred to as critically damped systems because it is critical they function exactly as designed or the system runs the risk of getting out of control and either destroying itself or stops functioning. Let's consider what would happen if our thermostat was designed and set to control the household temperature at 20 degrees +/- 0.1 degree. The furnace would shut on and off so frequently that the various systems of the furnace would be overworked and the furnace would quickly break; such systems are referred to as over damped systems and they are not desirable in most cases for regular performance. If the thermostat were set to switch the furnace on or off at 20 degrees either side of the setting, humans would not be particularly comfortable or happy and they would soon get rid of it and buy something more suitable. In the case of a furnace the furnace system probably would not fail but perhaps the fine wines and beers we are making at home would fail because the temperature would get too hot or cold for the organisms to live.

The wine and beer fermenting systems would be under damped, would fail; the organisms would die. Our fermentation ecosystem would cease to operate and no desired product would ensue; we might find that a different group of species would take over where the yeast left off; however, just generating *any* ecosystem was not our goal – we wanted a specific ecosystem and nature simply won't conform to human wants without a lot of human effort. We want our systems to be critically damped, timely, failure free and as productive as possible. Just in case our system might fail at a bad time we might build in some redundancy; we might have an extra furnace or we could parallel wire in an extra thermostat that changes in tandem with the first one. That way if one system fails we have some back-up and our risk will be reduced.

It strikes me that there is a considerable similarity between natural ecosystems and our critically damped thermostat/furnace system, only nature has built in a higher level of system redundancy. Normally the chemical and physical environment of an area of Earth stays pretty well within a predictable range and normally the plants and animals that have been able to contend with this range are able to survive there. If the changes in the environment are slow enough, the errors in the genetic code will allow some of the species to adapt. If the changes are too rapid, then the immense array of species that form the biotic potential of the area will kick in. Those that can tolerate the new conditions will thrive and replace those than cannot. The biogeochemical cycles are reestablished and the ecosystem restabilizes. Earth's system redundancy is so immense that it is immeasurable and has such a long history of successful functioning that it is inconceivable that it will fail; at least until the sun flickers and dies or explodes and wipes out life – at least on Earth. Maybe, just maybe, somewhere else in the universe. . . .

Humans have developed a lot of environmental control devices to let them modify what nature normally delivers; furnaces, refrigeration, buildings, insulation, machines, electricity and other types of energy, waste disposal systems, water systems, food systems, and transportation. They have also developed innumerable devices to control our environmental control devices; valves, switches and gates that operate pneumatically, hydraulically, electrically, electronically or with the help of computers and their programs. All are available with different precision levels, performance standards, failure rates and they can be in digital or analog form. The diversity of products in these areas is immense and if we don't have the device we will see if we can develop it. Desire and necessity are the mothers of invention. Regardless, we have developed a lot of devices to modify our environment, we know a lot about the organisms we wish to produce and we have a lot of knowledge of how to put these all together to form highly controlled systems.

Let's design and build a human controlled ecosystem to produce an industrial scale biological product. Let's produce chickens and eggs. Now, we remember, foxes like hens as much as rabbits like carrots but since we are controlling the system we will simply keep the foxes out. No worries. Roosters also seem to like hens and on rare occasions the likes are reciprocal but if we never let our hens know that roosters exist it probably won't trouble them much as long as they have lots of good food, and are kept comfortable and safe. Besides, we know that roosters don't lay eggs and when they're near hens the meat gets tough and humans don't like tough meat if tender meat is available for the same price. We'll keep roosters out of our ecosystem and let the chick supplier deal with the rooster problem. They probably don't need many roosters either because one rooster goes a long way in the chick raising business. Fifty-fifty odds on sexual selection don't bode well for the future of young cockerels unless someone can find a use for them – maybe someone will be interested in producing tender young cockerel meat. We are primarily interested in eggs. We'll forget about the cockerels; not our problem.

Now it hardly seems kind to keep chickens penned up if they have had the experience of running free with the sun and blue sky over their heads and the wonderful grass, grub and other things in a complex and diverse ecosystem – even if the foxes might get them and they have a hard time getting enough food and water. We will make sure they never experience the big world and if they don't experience it they can't miss it. That is the way organisms work; they are a product of the experiences they have and the environment they live in. If the environment is simple, the experiences are simple and the expectations beyond meeting basic needs are few to non-existent; that is of course if chickens are capable of having expectations. Just to be kind and err on the side of safety we'll assume they are capable of having expectations beyond their basic needs of food, water, health and sanitation; we'll keep their experience small. It also serves us well because chickens with unmet needs probably won't lay as many eggs; stress reduces egg-laying efficiencies and we are primarily interested in eggs.

We'll do everything we can to keep them comfortable; food, water, absence of distress, health and sanitation and they won't have to worry about a thing – as long as they keep laying eggs. It is in our interest to keep their environment this way because they will lay more eggs and waste little energy walking about and worrying about how they will get their basic needs met. Besides, too much walking wastes energy and the meat gets tougher in case we decide to sell them for food later. Certainly we want to keep them healthy and free from disease because if one gets a disease it might spread quickly to the others in such a small, highly populated place; that would neither be good for the chickens nor for us. We should also work out how to deliver the food and water and health needs and how to take away the waste and the eggs without disturbing and upsetting the hens; we'll work out the inputs and the outputs of our ecosystem and how best they can be delivered. That way we can keep the hens distress-free and perhaps save labor and costs in the process. It may take a little more thought and financial capital to begin with but we'll check out the economics through a careful cost/benefit analysis. Automation might just make sense for the chickens and for us but we had better build in some redundancy in our delivery and removal systems because we don't want any critical system to fail – that would not be good for the chickens or for us.

So we now have our general design specifications for our industrial egg production ecosystem. Lots of chickens, smallest space possible consistent with meeting chicken needs, balanced diet of food and water consistent with the best production of eggs, the best temperature, air quality and light for the chickens, good health care, good sanitation, stress-free environment, freedom from disease causing organisms, system redundancy and backup to prevent rapid efficient method to collect the eggs. Since chickens only live for a certain time and lay eggs for a shorter period than their life we had better partition the building into different areas so the old chickens can be replaced by young chickens with the whole chicken age spectrum represented. We don't want to shut down our operation and wait for the chicks to mature and start laying. We also want to have a continuous supply of old hen meat for our market. Continuous production of eggs and meat.

The building does not have to be pretty but it must be of high quality and not prone to serious failure; chickens are not into aesthetics and we will keep it well away from humans who might find the ecosystem offensive. Perhaps we should put some surveillance systems into place just in case some humans don't like our operation and try to disrupt it.

We need a stable ecosystem with all the sub-systems critically damped and working perfectly in unison all the time; that is the goal. We want the kind of stable environment in which our highly specialized orchid might have evolved; hopefully even more stable and unchanging, just less diverse and with less competition for resources.

Before we get down to designing and building the environment for the ecosystem and sorting out all the control issues we had better take a close look at the inputs and outputs of our planned ecosystem. Where will we get the food, water and health care products, how much will they cost and can we get them reliably delivered just in time to reduce cash flow, storage space and handling requirements? Where and how will the outputs from our system go: wastewater, feces, eggs and meat? The old hens are really a by-product of egg production, just like the feces and wastewater. Maybe all three can be reprocessed or perhaps we can find a market for them rather than pay to dispose of them. Do the future egg and meat markets look better or worse? What about delivery distances, prices, optimal egg characteristics and what are our chances of capturing a viable market share? How many eggs do we think we can sell and consequently how many chickens do we need to produce the eggs and do they meet our projected economies of scale? Perhaps since the chickens will be in an electrically lighted environment that is controllable we might be able to segment the barn so that the lighting varies according to our production and labor needs and have them producing continually day and night. Hundreds of other questions must be asked and addressed if we want our industrial biological production system to be stable and financially successful. Both the hens and we lose if we miss any important questions.

We do not need to carry our ecosystem design any further because the direction is clear and we have innumerable options for the construction of the ecosystem. We must work with our various consultants to sort out the details of the design, construction, marketing and economics. If we get our ecosystem design correct we will have developed the ultimate ecosystem with outstanding performance that takes much less space and uses exactly no more resources than necessary.

Such is the nature of industrial scale biological production. It is efficient, effective, and productive and hopefully it will be profitable until someone else comes along with a better plan that does the same thing with less wastage at a lower cost. Our system uses little space relative to a free-range chicken operation even when one considers that our facility relies on getting resources outside the system for inputs; these also are produced through intensive production; chicken feed, energy and chicks. Less space required to meet human wants in turn leaves more space and other resources on Earth for the other plants and animals.

Intensive operations like this one that meet all of our different human needs and wants could, if we humans wish, allow us to declare more parks and conservation areas on our land base. It could also increase our ability to use the land for other desired purposes – whatever we decide.

We must decide though. Each individual and country must have policies concerning land and resource use. If we decide incorrectly we may lose at a global scale to some other country that takes a different decision and follows a different path. Humans compete with each other for resources; intra-specific competition is a fact of life.

Preferential top carnivores have the most choice because when the preferred meat is not available they can be bottom feeders. Recall our discussion of the sophisticated human party we discussed a few chapters ago; 'plonk' works if fine rare French wines are not available. One can live on lentils and beans if he or she must.

On Resources

The notion of RESOURCE is really a very interesting one. For all life, a resource is simply something that is required to meet the organism's basic needs of food, water, shelter, security and reproduction. Reproduction is an optional requirement for the individual but not for the species. Some individuals must reproduce or the species disappears.

For plants the required resources are sunlight, water, air, soil, and space and these must contain the appropriate balance of properties and content; nutrients; water, carbon dioxide, oxygen, light, temperature and moisture. For animals the needs are similar only the animals get their food ultimately from the plants so the plants must contain the appropriate balance of properties; nutrients and palatability. Animals don't really need carbon dioxide that much but they have evolved to deal with it, in fact they require it to make other resources available. In fact all organisms have evolved or adapted to dealing with whatever is available in a particular area and if their needs are not met in a particular area they simply are not there. How they get these resources is programmed in their genes and they don't have a lot of options on the extraction and use method. One could quibble with my claim in a legalistic way and say organisms, particularly some animals, do make choices; they choose where to build a nest, establish their home range, whether to eat a banana or a papaya and I'll allow that I might be simplifying things a wee bit but I would be most surprised if many people would make the claim that animals have a lot of choice. If they do make choices I would be even more surprised if anyone would claim the choices are made freely, deliberately and with forethought.

All organisms need more or less the same things to meet their basic needs and seek little, if anything, beyond meeting these. The needs of each species are pretty well defined and do not vary a lot within a species. There is little if any innovation and variation in the collecting strategies of each species and given the limited space and resources on Earth and the large number of organisms, the competition for these resources is fierce. The resource war is waged in a small part of the planet; a few meters under the earth or water to a few hundred meters above the ground. Most organisms use only this bit of the earth to get their resources. Each has a particular range of environments and spaces it can meet these needs in for the resource needs are fixed. All organisms are the same in this with one exception; humans. The resources for humans are not fixed.

Humans are the exception among all organisms in the resource game for in humans the concept of 'resource' is highly diverse, highly variable and transcends meeting basic needs. Humans look at all of nature, think about it, study it, dream about it, ask themselves and each other questions about nature and provide answers to their questions. Humans debate their environment and decide whether it is either worth worshiping or whether it can be used. Without highly elaborate communication skills and the ability to form concepts and know we are forming concepts we would not be an exception; we would be out with the rest of the species competing ferociously in the regular resource acquisition game and we would be competing in the same narrow space of our planet.

Space is a serious resource limitation on earth and suitable space to meet an organism's needs is even more limited. It strikes me that a lot of nature's successful genetic accidents have allowed organisms to use spaces and resources that were not vital to other organisms at any particular point in time. Genetic errors and chance allowed the definition of resource to change and any organism that experienced such a change benefited until other organisms managed to develop methods of using the same resource or the original organisms collectively experienced enough genetic errors that other species evolved from them and began to compete. When only one species is seeking a particular resource then all competition for that resource is within the domain of that species; only intra-specific competition exists. Put another way; if a species can get by with less space, fewer resources or resources unused by other species then the other species have more resources available to them and the original species has an abundance.

Bunnies like carrots and if a carrot resource is available to bunnies they will eat it in due course; coyotes don't like carrots so they are not a resource to coyotes. On the other hand coyotes like bunnies.... Now if bunnies could learn to like coyotes and coyotes could learn to like carrots the resource competition and allocation game would be quite different. Thank goodness that carrots have different resource needs than either bunnies or coyotes or all three would have a tougher resource allocation problem.

Resource definition, use, allocation and efficiency are important and suitable space is perhaps the most limiting of all resources. The less space a species uses to meet its resource needs and the more those needs are different from other species the more space and resources are left for the other species. Such is the magic of humans – we have been dealt a different hand in the game of life; we have learned to do and pursue many things differently than all other species.

It is exceptionally difficult, perhaps impossible to define what a resource is to humans because it is always changing as our knowledge, understanding, technology and values change. It changes as we look at nature and imagine, consider options and actively change the Earth. Our ability to do these things sets us apart from the other species. Let's consider some things we consider resources that other species don't.

No other species actively pursues, processes, concentrates and competes for gold, tin, aluminum, silver, lead, zinc, titanium, platinum, iron, copper, arsenic, silicon or mercury. No other species actively pursues and processes, wax, rubber, oil, gas, wind, hydrogen, uranium or plutonium. No other species actively manages the land by tilling, spacing, planting and replanting corn, rice, cotton, peanuts, hemp or marijuana. Certainly no other species actively pursues concepts, facts, theory or education and certainly no other species actively imagines and realizes ways to convert these into things we need or want. No other species considers these things as resources in the same way. No other species has a concept of resource substitution and conversion.

Trees can be made into heat and light as can oil, gas, rivers, oceans, uranium and wind but they can also be made into other useful things. Trees to boards, paper, clothes, toys, houses, umbrellas; oil or gas into fertilizer, plastics, clothes, boats, planes, guns. No other species has used these primary resources and combined them with other primary resources to make secondary resources which in turn can be used to alter the primary resources to make tertiary, quaternary resources and so on. Stone and earth to make dams to make lakes and then dams, when combined with the resources to make turbines, generate electricity; electricity to light, heat and so on. All these in turn can be used to convert other resources to yet other products that we need or want – or think we need or want.

Just consider what resources are required to make a building, a city, a car or; coffins, computers, mines, airplanes or drugs. Consider what is involved and needed to manage a forest, produce carrots, chickens, wheat, electricity, oil, a painting, a symphony or money. Think about what is required to keep other organisms at bay and allow us to control nature and manage it to produce things we want. Think on the numerous agreements among people and the various resources of capital, money and time and you will be well on the way to defining what a resource is to humans. I won't expand the book by analyzing various other processes –

we have looked at some of the processes involved in natural resource production in ecosystems in earlier chapters. A resource to humans is whatever humans define a resource as. But let's not stop there in our resource definition . . . let's continue.

No other species pursues things of beauty simply because it thinks they are beautiful; diamonds, rubies, emeralds, rings, clothes, cars, yachts, parks, nature reserves, forests, grasslands, mountains or other species. Other species don't put products together that combine function and beauty to produce a beautiful product that optimizes form and function. If resources are combined and appreciated by other organisms they don't change over time based upon whimsy or values. Beauty is a human resource as are honesty, truth and kindness along with a host of other uniquely human values such as the concept of the public good.

If anyone has doubts about classifying values such as goodness, truth, beauty, honesty, fairness, trust and kindness as resources they would soon change their mind if they lived in a human society where these did not prevail. The absence of these in a society makes it a mean society to live in. In the absence of these values among humans, the environment becomes virtually uninhabitable. If you lived in one of these societies you would soon classify the above as resources and you would probably add a lot more to your value based resource list: laws, sharing, caring, loving, nurturing, benevolence, freedom, peace, acceptance, fairness, justice. . . .

Perhaps there is yet another level of human resources from which all these resources are derived. Perhaps it is our ability to ask questions. Perhaps the words: Who? What? When? Where? How? and Why? are to be numbered among our great resources. But they, in and of themselves, are not enough for in order to make these resources useful we must be able to ponder the questions and get answers. After that we must consider the answers in terms of their ability to help us manipulate all the other resources; primary, secondary, tertiary, values, morals. Finally we must call upon three other resources; the ability to plan, make decisions, and implement our plans. Let's examine.

On Evaluating Options: Decisions and Communication

The ability to evaluate options, take decisions and implement plans are critical human resources and all three rely exceedingly heavily on communication; visual, written, verbal and aural. In fact so critical to the human endeavor are these communication resources, they merit special consideration because they are a source of most human problems – and successes.

Sometimes the communication is so unbalanced, so fraught with error of meaning or so downright deceitful and manipulative that information is miscommunicated, misused, distorted and manipulated to suit particular ends. The ends of the communicator are not always clear because humans are exceedingly complex and one must always be wary of anything they read, hear, or see. They must test the quality of the communication and the information they receive by whatever means they receive it. Usually information is collected for specific purposes and if it is used beyond those purposes it might not be relevant or even useful – in fact it might be detrimental to the recipient.

Effective communication has for millennia been recognized as a problem that has been the focus of a great amount of careful thought and analysis. Proper and correct communication is an exceptionally difficult task even if honest communication is the goal and deceit and self-interest are ruled out. It is even difficult when all parties, the communicators and the recipients of the communications, have the same goal – forthright information transfer, good decisions, the public good, disaster prevention, conveying feelings and emotions and so on.

I have had the great good fortune for many years to share a table in our Faculty Club at the University of Calgary with a number of fine colleagues. We meet frequently over lunch and refreshments and we rarely socialize beyond this context. None of us have exactly the same academic interests for we are mostly from different departments of the university. None of us exercise any control over each other and if we could, we would not because it would be a breach of trust. The unwritten rules of the table and its communication would be broken and the group would be disrupted – trust would disappear and communication would become dishonest. That has never happened. We level each other psychologically, challenge each other, educate each other and help each other. We recognize our differences; our weaknesses, imbalances and strengths. We joke and laugh a lot – we are not 'politically correct' in modern parlance. We chide each other mercilessly but pain is never the goal. The goal is always honest exploration of ideas and the challenging of each other's ideas and values. We ruffle each other's feathers a bit but that is deliberate because we test and push each other but never so much that stability of the overall table, the group, is disrupted. Such places are rare; they are to be found, among many other places, in the coffee houses of Austria, the street cafés of Paris and the pubs of Britain. For some reason these places are almost always gender segregated; males with males, females with females. When a female joins our table the tone of the table changes; it is still pleasant but it is different. Why it is different I don't know and in this book I don't intend to pursue the reasons; respect, fear, discomfort, courtesy, embarrassment – who knows. It is a special place of communication and good communication is hard to come by. In the sociological literature they are called "Third Places."

Plato, in the *Phaedrus*, dealt with communication a great deal. We best persuade men, he noted, not when we intend to persuade men, but when we intend to uncover the truth; honesty is more important than cleverness and substance is more important than ornamentation. Plato carries his analysis a bit further and suggests that a speech is persuasive only to the extent that it resembles the truth. Plato is also a keen enthusiast of speech as a means of uncovering the truth. He suggests that *dialectic* – the conversation between people – is more effective than *rhetoric* – the art of presentation – in uncovering the truth. A conversation consists of statements made, responses offered, and responses given to the responses. Conversations grow naturally and the findings are not known until they arrive – there is no preconceived conclusion. In contrast, in a presentation the presenter already knows what points they wish to make and the statements of a presentation are arranged in a calculated way such that they will lead to the predetermined conclusion.

As I reflect on my career and my dealings with students, colleagues and various professionals and regulators in the areas of applied ecology and resource management I have concluded that there is huge merit in Plato's observations.

When one designs an experiment they must be scrupulous in ensuring their design tests an hypothesis, a statement whose validity is not known, and not a conclusion. If one designs their experiment to test a conclusion, important questions will be omitted even when the best of intentions prevail. If, for example, one wishes to determine the effects on river ecosystems of a pipeline that crosses a river by going underneath it there will be a tendency to conclude that since the pipeline will not directly alter the systems the effects of the crossing will be minimal to non-existent. This tendency may further lead the questioner to design a fairly simple experiment and ignore the possibility that during the process of drilling the hole for the pipe, the hole might collapse and indeed the ecosystems would be affected. It might also lead the questioner to ignore the possibility that chemicals harmful to the ecosystems might be required to lubricate the drill bit and should these be inadvertently spilled the ecosystems would be again affected. To be comprehensive in testing the pipeline problems there must be a series of separate experiments each designed to address part of the bigger question of the pipeline crossing and its effects on the ecosystems. The scientific method helps reduce these problems because it requires that the researcher establish two hypotheses; a null hypothesis that is tested and an alternate hypothesis that is not tested. In the river crossing problem one might establish their null hypothesis as 'The pipeline will have effects on the ecosystems of the river' and their alternate hypothesis as 'The pipeline will not have an effect on the ecosystems of the river.' Only after exhaustively testing the null hypothesis and being forced to reject it is one reasonably safe to accept the alternative hypothesis and conclude that the crossing will probably have no effects. The word 'probably' in the last sentence is important because certainty is not to be found in the future – science is fairly good at providing a sense of probability but not absolute certainty. Any number of unforeseen events could happen, for placing a pipe underneath a river is a complex process. Thoroughly modeling the event and then testing the model is important.

Discussion with colleagues knowledgeable about a particular range of topics is helpful. If our pipeline experimenter had laid out her problem to colleagues knowledgeable about pipeline crossings methods and requirements and ecosystems and allowed the conversation to evolve naturally she probably would have developed a much more careful and thorough experimental design than if she chose to design it without discussion. She would also benefit from discussing her data and results with others and later from having her report reviewed by impartial colleagues for she would receive the input of fresh unbiased eyes and in turn have a greater degree of confidence in the validity of her work. Conversations in coffee rooms, classrooms, faculty clubs and offices are important.

The previous discussion is not intended to discredit rhetoric and presentation but rather to illustrate the benefits of dialectic. Rhetoric does have its place. It can be useful in presenting tested and known facts and concepts in a relatively quick and efficient way. In a classroom it is possible to lecture and convey a huge amount of information to a large number of students in a short time. The presenter should select the information to be presented carefully and convey it as clearly and precisely as possible. He should also attempt not to persuade the students to one particular view or another but rather show the weaknesses and strengths as much as possible. What the student (or the recipient of a presentation if it is not a classroom context) does with the information after the presentation is equally as important, perhaps more so, as the lecture itself. The presentation information needs to be carefully considered and tested in the light of reason, experience and judgment. Insofar as a presenter honestly seeks balance and fairness in their presentation they are providing a great service to their audience; insofar as they attempt to persuade by selecting only that information that is in support of their point, they provide great disservice unless they clearly state there is contrary information and direct their audience to examine the other side of their arguments.

Good presentations often are entertaining, interesting and humorous for they keep the interest of the audience. Entertainment and humor must, however, be byproducts of the presentation and not form the main part. They are wonderful ornament but they must always be subservient to the substance.

Presentation is not restricted to the spoken word. The written word – reports, books, articles, theses – are forms of presentation and in many ways they can be even more dangerous and powerful than the spoken word for people have a tendency to believe the written word more readily. With the written word the author can carefully craft his arguments and be exceedingly persuasive through selectivity. Arguments can be constructed and conclusions presented as fact when indeed a careful analysis would show it to be only one possible arrangement of information.

In all communications there is considerable merit to being skeptical about the information one receives; a healthy dose of methodical doubt is beneficial as it helps clarify, refine and test the information received. Thoughtful people consider information before accepting it – they critically analyze their information to determine its merits. Let's examine information transfer. We will focus on the written word for it has the greatest limitations to the human endeavor (but it also requires more discipline than the spoken word).

The written word is the most limited for the writer only has letters, words, sentences, paragraphs, punctuation and rules of grammar to work with. In contrast, the spoken word in conversation is much richer. It relies heavily on non-verbal communication; the smile, frown, voice inflections, intonations, variation in modulation and loudness, appearance, a wink, tentative statements and so on. Insofar as a writer can capture the richness of the spoken word they are able to improve the quality of their communication and improve understanding. It is difficult to achieve this but one must try for it assists the reader greatly.

Grammar, or the arrangement of words according to rules, bears a huge burden in communication for it must make the relationship of words instantly and unmistakably clear. It defines how words relate and brings clarity. Insofar as it does not achieve this clarity, miscommunication will result.

In decision taking, miscommunication and errors in thinking often result in exceedingly bad judgments. The best defense against bad judgments is reason. We will examine some of the common errors in thinking because they are critical to good decision taking. They are critical to understanding nature and the human endeavor; in managing and changing nature in accord with our wishes, needs and wants.

It is critical to understand the difference between facts and opinions and how they relate to each other in making judgments. They are often confused. A fact can be verified while an opinion is an attitude toward facts; opinions are subjective. Opinions are drawn from facts and the more facts that support a particular opinion and the judgment concerning it, the more reliable that judgment is. In making judgments we must be concerned with the truth of facts and the validity of opinions. A fact is true only insofar as it conforms to reality and validity means well-grounded in evidence. Evidence consists of the testimony of the senses, testimony of witnesses and the testimony of authorities. The validity of a judgment then depends on the facts that support it and the ability of the person making the judgment to assemble the facts in accord with reality and probability. Let's take a simple statement concerning nature and examine it.

If someone says, "If that species becomes extinct there will be one less species in the world and if species continue to become extinct the diversity of species in the world will be reduced." On the surface the statement sounds reasonable and believable. What is fact and what is opinion in this statement? Certainly if a species becomes extinct it will no longer be alive on earth and if other species suffer the same fate they will not exist either – these are facts. On the other hand the claim that the diversity of species on earth will be reduced is an opinion and we do not know whether or not it is true – the evidence is that new species are evolving continuously on earth and there may have been more new species added than the number of species that became extinct. Facts and opinions have been confused.

Humans experience things and events in the particular yet the human mind has a tendency to form universal connections and make hasty generalizations concerning their experiences. Hasty generalizations can have a huge impact on judgments and decisions if they are not ferreted out and examined to determine how valid they are. Let us look at two examples.

Consider a pipeline company that has a great deal of experience in pipelining in southern Canada and the northern United States, excluding Alaska. Let's also assume the company has had an excellent operating record with few pipeline failures and ruptures. Let's further assume also that the pipeline company wishes to construct and operate a pipeline in the arctic. As evidence of its ability to do this successfully it claims that its successful past operations provide solid evidence of its ability for the proposed new undertaking. This claim is a hasty generalization because all of the particular experiences of the company are in quite different environmental conditions than those that prevail in the area of the proposed pipeline. The company has had no experience constructing and operating a pipeline in permafrost conditions and conditions of extreme temperature variation. Their generalization needs to be tested carefully to see if it applies in the new conditions.

Our second example will also be drawn from the pipelining field. Let's assume the pipeline company proposes to construct a pipeline across a piece of unbroken prairie grassland and objections are raised that it will have detrimental environmental effects on the grassland ecosystems; it will fragment the habitat and have negative effects on the plants and animals. There is much evidence to demonstrate that some species are affected negatively by habitat fragmentation ,however there is no evidence to show that all plants and animals will be negatively affected. In fact, evidence and reason suggest some species will benefit from the pipeline right-of-way and other species will suffer. The effects are species specific and in order for the claim to stand up, the evidence must be specific. A hasty generalization has been drawn from the evidence that shows some species are negatively affected; a universal conclusion has been reached. It behooves the proponent and the objector to provide further evidence in support of their claim that the effects are either acceptable or unacceptable in order for a sound judgment to be reached. Such judgments are particularly difficult because they involve matters of human values – is the grassland so valued by humans that it should not be disrupted? The plants and animals will either adapt to the environmental change caused by the pipeline or they will not. If the former ecosystem is re-established with time it is likely that the effects of the pipeline will only be temporary – time is important in ecosystem change as we have discussed in earlier chapters.

The relationship between causes and their effects is exceptionally hard to establish and in many situations is effectively impossible. Terribly bad environmental management or personal health decisions are often made when one assumes cause-effect relationships where they may or may not exist. We know that events have causes and that causes precede effects and we often have a tendency to assume a relationship without testing it to see if the relationship is valid, accidental or spurious. Scientific experimentation expends a great deal of effort in trying to establish cause-effect relationships. To establish these relationships many experiments, depending on what they are designed to test, attempt to control and hold constant all the variables except one, which is changed. The effects of changing that one variable on the subject of the experimentation are documented and attempts made to establish a cause-effect relationship.

To demonstrate my point we will consider an experiment to determine the effects of light intensity on a particular plant species. In order to do this a researcher would normally determine what resources the plant needs to grow and would provide these in such quantities that they would be uniformly available; neither in short supply nor abundant to the point they would hamper the growth of the plant. Light intensity would then be varied and the effects of the different light intensities on the growth of the plant recorded. If the experiment is successively repeated and the same results achieved with each experiment then a likely cause and effect relationship has been established. Subsequent experiments might hold light intensity constant and vary one of the other variables to determine what effects these had on the plant growth. With sufficient experimentation the effects of resource availability on the growth of the particular species would gradually be revealed. The interrelationships among the variables could then be established by multivariate analysis and a model of plant growth, and its relationship to the different variables, constructed. Establishing cause-effect relationships is slow, tedious work but it is exceptionally important in understanding our world. Let's look at a couple of other examples.

Consider a discussion between a very old lady and some considerably younger friends. The friends ask the old lady to what cause does she attribute her longevity. The old lady responds that she recently read in a newspaper about some research that showed 'free radicals' in one's diet were responsible for good health and that broccoli had an abundance of free radicals. Upon reflection on her past diet, she explains that she had always eaten a lot of broccoli and that she now attributes her longevity to broccoli consumption. The elderly lady's reasoning is after the fact and it is undoubtedly flawed. The human organism is just too complex, too variable, and longevity and health are just so poorly understood that it is simply impossible to establish such a simple cause-effect relationship with any validity.

We'll now tackle an even more complex cause-effect environmental problem – global climate change. Let's assume that we notice global climates have been warming and this concerns us because the seas might rise and flood a lot of areas where humans live and the humans will be displaced. Furthermore the warming climate may change the global ecosystems and increase the abundance of deserts that will result in lower food production and increased starvation. Someone notices the carbon dioxide concentrations in the atmosphere have increased as the earth's temperature warmed and establishes a correlation between the two. Now let's assume someone else notes that humans, through combustion of fossil fuels, emit a lot of carbon dioxide into the atmosphere and suggest that humans may be one of the causes of global warming. The news media then get hold of this carefully phrased conditional statement and headlines "Humans Cause of Global Warming" "Global Disaster likely due to Human Carbon Dioxide Emissions" are communicated to the public. Public fear and concern build and more facts and information are gathered and assembled to demonstrate the validity of the human caused global environmental warming claim. With each successive finding and news release the fear is increased and calls for action are made to stop humans from causing global warming. The rhetoric increases and all kinds of action are taken as a result. Groups call for stopping the production of certain energy types and governments are called upon to reduce the carbon dioxide emissions within their political jurisdiction The governments act when a large number of voters demand it so people will continue to vote for them. Self-interest prevails. The process continues and the stakes get higher and higher. Poorly established cause-effect relationship claims have been made, or assumed to exist, and the effects of these claims on the human endeavor have been extensive.

The initial observation that warming climates can be correlated with the carbon dioxide concentration in the atmosphere may well be true but a correlation does not a cause-effect relationship make. The observation that humans burn fuel and emit carbon dioxide into the atmosphere is irrefutably true however humans are not the only source of carbon dioxide emissions into the atmosphere. No one knows how much carbon dioxide is emitted into the atmosphere by total global processes and no one knows how elaborate the feedback mechanisms in the global carbon cycle are and how nature will deal with them. Without knowledge of how much carbon dioxide is emitted into the atmosphere it is impossible to determine how significant the human contribution is to the overall warming process. Elaborate global models are built, based on sets of assumptions, and these are used to support or refute a particular action. Some claim, based upon their data set, that humans have no impact and we don't have to worry about disaster and we can keep on emitting carbon dioxide heedlessly. Yet others claim that unless we drastically reduce our emissions global disaster is inevitable.

The above scenario is predicated upon an observed correlation that does not have a proven cause-effect relationship – only a possible cause-effect relationship. A possible relationship and an important one that should be of considerable concern to humans has been established. What actions should be taken in cases such as this? The costs of error in judgment are enormous regardless of which conclusion we reach; we are firmly planted on the horns of the proverbial dilemma. In situations of uncertainty, prudent people act prudently- perhaps we can disempale ourselves from the horns and sit squarely on the head of the dilemma that holds the horns. There is enough validity to our concern that we should be cautious and try to reduce our carbon dioxide emissions. There is also enough uncertainty to suggest extreme measures are inappropriate, until such time as we have enough information to demonstrate that extreme action is required.

One must be exceedingly watchful for cause-effect relationship problems and post hoc reasoning when they make judgments and take action. Unless they are watchful they may well jump from the frying pan into the fire and take a lot of other creatures with them. The validity of cause-effect relationships is usually 'outed' in the long run. The Inquisition and witch burning ended; Louis Pasteur convinced us that flies do not spontaneously generate from rancid meat.

The use of analogies in exploring knowledge and seeking the truth, the conformity of our knowledge with reality, can be useful and powerful but they must be used with caution because analogies between things or processes always break down – when they break down and no longer apply they can cause problems. I drew analogies in earlier chapters when I noted that growing trees in plantation forestry is analogous with growing carrots in intensive agricultural operations. I also drew an analogy when comparing the management of native grasslands to produce beef and the management of extensive forestry operations for tree production. These analogies are true to a point, there are many ways in which each is parallel, but they certainly are not the same. One must be exceedingly careful not to carry an analogy too far.

If, in trying to determine the possible effects of a pipeline built through a forest ecosystem we draw a parallel between a pipeline and a road it can be quite useful as a conceptual model to elaborate and refine our thinking on both the differences and similarities. Both pipelines and roads are long linear facilities and the trees must be cut down and the forest bifurcated in order to construct them. Both tend to be fairly permanent landscape features. The physiognomy of the forest will be altered in many similar ways and in many ways the effects on various aspects of the ecosystems will be similar. The analogy breaks down when one considers that pipelines are generally constructed and then actively revegetated or, in the absence of active re-vegetation, left alone to be recolonized by species that arrive there as a result of the area's biotic potential; providing of course the organisms that recolonize there do not cause problems in the maintenance and operation of the pipeline. Roads, in contrast, are usually kept free of vegetation and they are continually disturbed and used. Because of these differences roads and pipelines are quite different in their effects on forest ecosystems. The analogy can be useful in exploring both the similarities and the differences between or among things, however one must recognize when it is essential to switch from discussing similarities and begin discussing differences.

A word gets its meaning from a social contract; an agreement among people that when it is used it will mean a certain thing. Often a particular word has different meanings in different contexts and if, in the context of a particular communication, the meaning of the word shifts, a problem known as equivocation arises. Equivocation can be a tremendous source of poor decision-making and it can also be terribly frustrating. Words such as wrong and right are notorious sources of equivocation as are other words that relate to human values; integrity, stability health, wealth, care, poverty and so on. Let's examine equivocation.

Suppose someone says, "It would be right to log this area because it will create employment and reduce human suffering." What does "right" mean? Does it mean necessary? Moral? Good? If you disagree with the statement would you then be inclined to argue that creating employment and alleviating human suffering are not desirable (assuming it has been demonstrated that employment and reduced suffering will be the consequence of the logging)? Some people might argue that it is necessary but not good to cut down the forest. Others might argue that right simply means necessary and yet others might argue that right means moral. If all these people were in the same room debating whether or not to cut down the forest the debate would be full of confusion and misunderstanding because each group of people would be discussing different questions and each would muster different arguments. The conversation would shift from one meaning to the other and no consensus could be reached.

Is cutting down the forest "good"?

Is cutting down the forest "necessary societally"?

Is cutting down the forest "morally sound"?

It is essential for a productive discussion that a common definition and understanding prevail. Many conflicts in environmental and resource management decision taking arise from equivocation and no one feels as though they have been heard.

What do terms such as *ecosystem integrity* or *global biodiversity* really mean? Does ecosystem integrity mean unchanging? Stable? Conforming to the state of the system at some particular time? Do ecosystems even possess integrity or do all ecosystems have integrity? What do we really mean when we talk about 'preserving global biodiversity?' Do we mean all species should be valued and prevented from becoming extinct? Would people who wish to preserve global biodiversity argue that we should not eliminate the smallpox virus or malaria? Anthrax? Often words or concepts such as these bring confusion to decision taking and evaluation unless we can establish a common platform of meaning; words and concepts to be of particular value must have a common meaning and this involves a social contract among those using the words or phrases.

People think and analyze things in two different ways; we either apply a general rule to specific cases or we examine specific cases to formulate a general rule. Reasoning from the general to the specific is called deductive reasoning and reasoning from the specific to the general is called inductive reasoning. In order for a general rule to apply to a particular situation there must have been enough specific cases examined to verify the applicability of the general rule otherwise errors and incorrect decisions will be the result. Conversely, in order to develop a general rule, enough specific cases must also be examined and found to be consistent otherwise no general rule can be developed. We get our fundamental sensory-derived evidence about the world in the specific and we then aggregate and classify this specific evidence into the general categories to help us comprehend the complexity of the world. These general categories are useful in making predictions.

If people are not persuaded that a general rule is valid, no amount of argument and logic based on that general rule will persuade those who do not accept the general rule. If they reject the general rule because, from their experience or reason, they do not think it is valid then further evidence and testing might convince them. If they reject the rule because they believe it is invalid, no amount of evidence will convince them otherwise for they are acting on belief and not reason. Beliefs are non-debatable unless the believer is prepared to set their belief aside, at least temporarily, and examine it. Fundamental beliefs are not necessarily reasonable; they are acts of faith.

If a person states "I am opposed to this project because I believe it is immoral," there is no way to persuade that person the project is moral unless the communicators are prepared to discuss morality. Likewise if someone states "I am opposed to killing these animals because I believe they have rights and their rights are more important than the rights of humans to have the animal as food," there is no possible way to persuade them there is merit to the slaughter. Discussions between people that involve different fundamental beliefs will never be resolved as long as the beliefs hold firm; reason, deductive or inductive, is simply ineffective.

The classifications of things or events can cause problems in revealing the truth when they are either faulty or inadequate. To be useful classifications must be appropriately exhaustive and the categories must be mutually exclusive. If one item can't find a home in one of the categories then it is clear the classification is not exhaustive; if the classification of that object is important to the matter under consideration then the classification is inadequate.

Unless classifications are exceedingly simple and the individuals to be classified fairly uniform it is nearly impossible to develop a perfect classification. It seems to me that the classification of nature, because nature is exceedingly variable, diverse and complex, will always be imperfect at best. The classification of natural objects and processes will likely never be fully comprehensive for in nature there are so many intermediate forms; classifications are generalizations of nearly continuous variation. The lack of perfect classifications should not be particularly troubling because generally one can develop a classification that is good enough for the purpose at hand; the classification however must be 'good enough.'

I recall one situation where an individual showed me a map on which he had carefully drafted all the linear facilities he could locate – roads, oil pipelines, gas pipelines, seismic explorations lines. He used the map to try and persuade me that all the linear disruptions so fragmented the ecosystems that the ecosystems no longer had any integrity. He made no effort to differentiate and classify the different types of linear facilities, explain how they might have different effects on the ecosystems nor did he attempt to explain how the different distances between the facilities might affect the ecosystems. He failed to tell me what he meant by ecological integrity. I was simply presented with a map as solid evidence of the impact one further linear facility would have on the ecosystems and left to draw my own conclusions. I did not know how to use the information for the lines on the map were not differentiated nor were they appropriately classified. To this day that event troubles me because I believe he was on the right path to a making an important point. When questioned on this matter he was unable to elaborate on and explain the significance of his map. Regrettably all of his fine work went for naught – I was left frustrated with his evidence and he was left bewildered by my inability to understand his point. The context of the situation prevented me from ever resolving the issue.

In many communications one is faced with arguments based on false dilemmas. These often arise when the arguer draws simple extremes in their arguments. They are generally fairly easy to recognize and quash or put in perspective. Suppose someone argues that a proposed road must not be built because if built, it will result in the destruction of populations of a rare and endangered plant species. The person has drawn two simple extremes and left no room for a middle ground; we are asked to choose one or the other – build the road and destroy the plants or not build the road and save the plants. The person has not allowed that the road could be relocated to avoid the populations nor has he acknowledged that the populations could be either moved to another location or the plants could be bred to expand their numbers and reestablished in many areas thus reducing their rarity.

Over-precision in arguments is often a problem that can derail analysis and decision taking. Words and concepts can get torn apart and rendered useless as a result of over analysis and a discussion derailed. Sometimes this is referred to as hair-splitting, legalese or logic-chopping. Let's suppose that as part of a revegetation project a proposal is put forward to plant native species in the area because native species are desired. It is proposed to use the seed from native species that have been genetically selected and grown with agricultural techniques because the cost is attractive. The logic-chopper might argue that because they are genetically selected they are no longer native. Let's then suppose our logic-chopper carries the day with the soundness of his arguments and the proposal is modified. A decision is taken to reseed only with seed from individuals of the species that have not been genetically modified. After some searching, genetically unmodified field grown native seed is available in Alberta, a considerable distance from the revegetation site. Our logic-chopper could then complain that the seed, though native and of the same species, is probably genetically different from that which is closer to the target site. Additionally, the introduced native seed might crossbreed with the local plants of the species and the genetics of the local population would be altered and made less native. The logic-chopper could continue in his pursuits and next insist the seed must be wild-collected from the local area. He can continue endlessly to subsequent levels of precision: only from plant communities with similar soil types; only if the seed is collected from various periods through out the growing season and blended so the whole spectrum of specie's local genetic stock is represented (genetic variation can result because individual plants of a species produce viable seed at various times throughout the growing season). This is a classic case of over-precision and it can take many forms in the environmental management arena.

Often appeals to authority are used because an appellant assumes his assertion will carry more weight if authority is cited. The assumption is correct but only if; the specific expert is identified, their credentials established, they are recognized by their peers as sound in their knowledge in the relevant area and they do not have a vested interest. It is inadequate to say, "experts agree ..." or "climatologists have found. ..." One must ask: Which experts? Which climatologists? All climatologists? And so on....

One comes across answers in conversation that commit the fallacy of begging the question – giving the conclusion as a reason for the conclusion. When someone is asked why a particular species should be protected and they answer: "If we don't protect this species it will have no protection" is begging the question. Nothing is gained from this interchange unless, of course, the reasoning and the answer are accepted as valid and used to make a decision or take action.

Occasionally one comes upon arguments related to resource management and development that are based upon popular appeal – '...the vast majority of people feel that this area should protected (or this mine should be approved) therefore a decision should be made to protect it.' Just because the vast majority believes or feels something should be done does not provide any evidence upon which to make a solid decision based upon reason. The vast majority may have formulated their position from inadequate evidence, faulty analyses or they may have been persuaded by campaigning or advertising.

Not infrequently in issues related to taking resource management decisions one will be witness to a personal attack. Personal attacks are not really faulty reasoning, they are simply 'bad behavior' and they are designed to defame the character of an opponent and thus discredit his evidence or opinions. Sarcasm, recriminations, abuse or personal affronts are common methods. In situations where claims about a person's character are used as justification for accepting or rejecting their evidence it is entirely appropriate to examine and test these claims; the validity of their evidence is based upon the validity of the character assessment.

When one assesses arguments of a person it is also important to consider the motives or possible bias or vested interest of the person making the arguments. If the person has a vested interest or is biased in one direction or another, or has a particular motive to slant their view, they may be unable to be fully objective in presenting their arguments. It is only human to at least have a slight tendency to work things in favor of one's particular position. It is important to note that positions of vested interest or bias do not necessarily signify the evidence given by a person in such a position will automatically be biased. The evidence and arguments of such people must be evaluated on their own merit through careful analysis and given weight accordingly.

Yes. Communication is very important in the human endeavor!

Options on Futures

I have chosen the heading Options on Futures for the final chapter of this book because it strikes me that the future of nature and of the human endeavor is a bit like the futures market of commerce. There is a certain element of chance and happenstance concerning tomorrow, or next week or next year – one can do nothing about this because chance is not predictable. For something to happen by chance, however, it must be within the realm of possibility, things that are not possible do not happen.

When you buy 'one beef' on the futures market you are betting that nature will continue to deliver beef from its ecosystems (of course with a bit of ecosystem manipulation by humans) and on a certain date you will receive forty thousand pounds of it to do with what you want; hopefully sell it to a slaughter house before it is delivered onto your front lawn for forty thousand pounds of beef on your front lawn will probably upset your neighbors. When you buy a beef you are betting that the sun will continue to shine, the grass will continue to grow, the cows will continue to eat the grass and humans will continue to eat cows. When you buy a beef you are also betting that by the time your futures date arrives, nature will not have been quite as kind to the beef industry as it was when you bought your option; if nature is less kind to the beef industry the price of your beef will have gone up and you will make a profit on its sale. If you did all your homework very carefully and made all the right assumptions about nature and cattle and took as much of risk out of the purchase as possible you will probably make a profit. The profit is wealth that can be stored or squandered as the case may be. If you save some of your profit to attend to tomorrow's needs you are wealthy because tomorrow's needs are assured should you still have needs tomorrow. If you don't have needs tomorrow then surely someone else will get your surplus.

If you don't have enough surplus to meet tomorrow's needs then you must use your hands, feet, knowledge, and with the assistance of human designed and made implements, you must gather tomorrow's needs from nature. If tomorrow's needs are not readily available you will do whatever you must to meet those needs; you will poach animals in a park or steal carrots from your neighbor if you must.

If you are unable to provide for tomorrow then you must either step away from the gaming table of life or you must rely on the stored wealth of others to tide you over until once again you are able to provide for yourself. If your society has enough accumulated and stored wealth its members will be able to assist you with tomorrow so you do not have to poach animals in the park or steal carrots from your neighbor. They will also, perhaps, help you with tomorrow's needs because they feel that you will be able to contribute to their well-being in the future when you get things sorted out and because they value your life and feel starvation is unacceptable.

If you receive the beneficence of society you must both recognize the assistance you have received and your obligation to make every attempt possible to again provide for your needs and contribute to the collective societal surplus so that others, if they need assistance, may call upon it. Insofar as you and other members of society build wealth and provide for tomorrow's needs, the society will do well and animals and carrots will not have to be poached or stolen. Insofar as societies wealth is increased because more wealth is stored, the society has more options to plan and consider how best to conduct itself in the future. Insofar as the social surplus is reduced because more take than give, the society becomes poorer and less able to meet its future needs. When a society can no longer meet the immediate needs of its members, the individuals must compete with each other and with other organisms in order to meet these needs or they must rely on the beneficence of other societies who have accumulated and stored wealth to assist them to provide for tomorrow.

If the wealth of all societies is reduced so that none remains, there will be no surplus for tomorrow and basic needs must be competed for in the environment today and every day. When immediate basic needs are not met individuals will do whatever they must to meet these basic needs. When the basic needs are unmet, the health of some individuals will decline in accord with their lack of resources. As individual health declines some individuals will be less able to compete with other organisms for the resources. They will be less able to reproduce and raise healthy offspring to replace themselves. Some will die. As fewer offspring are born and more individuals die, the population will decrease because the death rate will exceed the birth and survival rate. As the population declines the species will have less call on the resources it needs. The resources, if they are biological and renewable, will begin to increase and more resources will be available for the remaining individuals of the population and for the individuals of other remaining species. If the population declines beyond the demand it has for the resources, those renewable resources will increase in accord with demand places upon them. As more resources become available, health will increase and the birthrate will increase; the death rate will decline. The population will increase and as long as it does not use more resources than are available it will be able to store wealth if it so chooses. As stored wealth increases, there will be less care for tomorrow and the less fortunate individuals will again be able to receive the beneficence of the stored wealth, if the society so chooses to share. With less care for tomorrow, individuals in the population will have time to contemplate the world and gain more understanding about what it is comprised of, how it works and how it might be manipulated for whatever purposes the society chooses. Choice, within limits, is a very powerful human characteristic.

Human choices may either be based upon reason or not based upon reason. If they are not based upon reason they may be based upon belief, ignorance or indifference – if they are based upon belief they are not subject to reason unless the belief is questioned. If they are based upon ignorance or indifference they are neither made on the basis of reason nor belief.

Reason seeks the truth and truth is conformance with reality. Belief also seeks the truth but it is not ultimately subject to reason. Conformance of our knowledge with reality is important because it allows us to make choices and take decisions and actions that are based upon likely outcomes. Unless reality is fully understood, decisions and outcomes will never be fully predictable. Lack of predictability of an outcome leads to uncertainty of an outcome – the greater the uncertainly of an outcome, the greater the risk that an action resulting from a decision will not result in the desired outcome. Reason seeks knowledge that conforms to reality; actions based on conformance with reality produce more predictable outcomes and humans benefit from being able to predict the outcomes of their actions. Insofar as reason leads to a desired outcome from their particular decisions and actions, humans benefit from reason.

I have argued, hopefully reasonably, that insofar as humans are able to capture and use resources that other species do not use, or do not have access to, the other species will have more resources available for their use. If humans wish to maximize the resources available to other species, either because they need or value these other species, they must use their knowledge and skill to make choices that will lead to that outcome. We have demonstrated that we are able to meet our needs and wants and develop technology that allows us to use resources others do not use; from deep in the earth we draw oil, gas, uranium, iron and such like. We have demonstrated that, because the particular genetic code of our species allows us, we are able to; examine the past, assess the future with some confidence, think abstractly, store knowledge, evaluate options, fabricate items and alter existing ecosystems by choice in order to meet our own ends.

We have demonstrated that we are able to concentrate the production and storage of our resources so that they are produced or garnered from less space and with less wastage than if the same amount of resources were to be harvested from nature's unaltered bounty. We have demonstrated that we can use energy and technology to alter unsuitable environments and make them habitable when they otherwise would not be habitable by us. We have demonstrated that we can produce our energy and other needs from less space as our experimentation continues and our technology develops. We have demonstrated that we are able to meet our energy needs from progressively less space as we discover and use more concentrated energy sources; producing energy from trees takes more space than coal, coal more than oil and gas, oil and gas more than uranium. We have demonstrated that we can manipulate natural ecosystems by simplifying or changing them so that we can produce more of a particular biological need from less space: more grain is produced in a field than would grow in areas unaltered by humans; more chickens are produced in a intensive industrial agricultural operation than are produced in a more conventional chicken coop and a conventional chicken coop produces more chickens per unit area than we could gather from nature unaltered by humans (the feed for such intensive operations is also produced in less space with intensive plant production than nature would require). We have also demonstrated that we can manipulate and change the genetic code of other organisms so they can, with our assistance, produce more of what we want and from less space and less use of resources than the un-manipulated organisms would: genetically manipulated fruit trees produce more fruit, genetically manipulated grain produces more grain, genetically manipulated cows produce more beef. We have learned to control our own reproduction and the reproduction of other organisms to a fairly high degree.

We have learned, sometimes through hard lessons, that concentration and simplification of systems increases the risk of failure of those systems: the rapid combustion of a kilogram of wood is less risky than the rapid combustion of a kilogram of oil and the rapid combustion of a kilogram of oil less risky than rapid fission of a kilogram of uranium or plutonium; the invasion of a battery chicken operation, an intensely managed forest, or an intensely managed grain field, by a disease or pest causes much greater loss of resources that humans want than the invasion of a natural ecosystem by a similar pest – we have concentrated the resources these organisms use and they multiply in accord with the resources available to them.

We have learned, through experimentation and experience, various ways to reduce our failure risks due to concentration and simplification: we contain and control combustion; we keep out, eliminate or introduce organisms in our systems and we help or hinder these organisms in meeting their resource needs by adding or taking away resources in accord with our wishes and needs. We have learned to reduce the failure and risk of our concentrated systems by introducing redundancy into them; if one essential part of the system fails there is another part that will either pick up and meet the needed resources or deprive the system of resources. Irrigation supplies resource redundancy to our agricultural fields if the rains fail; automatic or manual opening and closing devices deprive or supply our systems with the needed resources in order for them to keep functioning or prevent them from functioning. As appropriate, we have learned to make our systems somewhat error free and error tolerant. We have not learned to make any of our systems either completely error free or error tolerant. Our collective experience suggests this is not possible. Planes will continue to crash; houses will continue to burn or fall down; the systems we construct will not always function the way we want or expect them to; crops will fail.

We have learned through experience, sometimes as a result of great tragedy that humans do not always act in the best interests of each other, or of themselves – even if they think they are so acting. We have learned to develop codes of conduct to control people's actions and we have learned to judge people's actions in accord with these codes – and take what we believe to be correct and just actions against those who do not comply with the codes. We have learned, usually in hindsight, that our codes of conduct and our laws are not always fair or just or moral and we have learned to eliminate or modify our laws so they are in accord with what we presently understand to be just, fair and moral. We have also learned how to reflect on our errors and apologize to the offended individuals and make recompense to the best of our ability. Recompense is not always possible and we have also learned that some people will seek recompense based upon our laws because it will increase their wealth at the expense of other members of society; people who are hurt sometimes wish to hurt back; even though hurting back does not always help these people to move on and live their lives as best they can or wish.

We have, through studying and considering nature and our place in it, learned that nature continually changes and the species abundance and distribution continually change as the environments those species live in change. We have also learned that those species unable to get their needs met in the new environment will be eliminated from that environment – if there are no suitable environments remaining for that species or if it cannot reach those environments it will become extinct. The biotic potential of an environment determines whether members of a species can reach that environment and find the necessary resources to survive; humans have a huge ability to alter the biotic potential of all environments; we do so inadvertently through many of our actions (consider the Suez or the Panama canals; consider also our many modes of transportation) and we may do so deliberately if we so choose.

We have learned through studying nature and our place in it that nature has always changed and adapted to the new conditions and will likely continue to change and adapt in the future; even though these changes may neither be in accord with our wishes nor our time-frames.

Many people have not yet learned to accept that we are ultimately subject to the laws of nature and that natural systems will continue to adapt and change whether or not we remain part of them. Natural processes always act.

Many people have not yet learned they always have some control in all the situations they find themselves in – they may not be able to alter the situation however they do have control over how they view that situation.

Many people have learned that finding meaning in life makes their life richer and many people have found it important to ask themselves "Why am I here?"

Many people have found, when they contemplate this question, it is important to them to choose to have faith – in whatever or whomever – because it adds meaning to their lives.

Many people have found that reason, regardless of how useful it may be in providing humans with resources and a better understanding of themselves and nature, does not provide all the answers.

AGMV Marquis

MEMBRE DE SCABRINI MEDIA

Québec, Canada
2004